# THE E-BOMB

# THE
# E-B●MB

How America's New

Directed Energy Weapons

Will Change the Way

Future Wars Will Be Fought

# DOUG BEASON,
# Ph.D.

**DA CAPO PRESS**
A Member of the Perseus Books Group

Designed by c. cairl design
Set in Minion by the Perseus Books Group

The Library of Congress has cataloged the hardcover edition as follows:

Beason, Doug.
   The E-bomb : how America's new directed energy weapons will change the way
future wars will be fought / Doug Beason.
      p. cm.
   Includes bibliographical references and index.
   ISBN-13: 978-0-306-81402-0 (hardcover : alk. paper)
   ISBN-10: 0-306-81402-1 (hardcover : alk. paper)    1. Directed-energy weapons.
2. Nonlethal weapons—United States.    I. Title.

UG486.5.B4345 2005
358'.39—dc22

2005013291

First Da Capo Press edition 2005
First Da Capo Press paperback edition 2006
ISBN-13: 978-0-306-81506-5 (pbk.)
ISBN-10: 0-306-81506-0 (pbk.)

Published by Da Capo Press
A Member of the Perseus Books Group
www.dacapopress.com

Da Capo Press books are available at special discounts for bulk purchases in the U.S.
by corporations, institutions, and other organizations. For more information, please
contact the Special Markets Department at the Perseus Books Group, 11 Cambridge
Center, Cambridge, MA 02142, or call (800) 255-1514 or (617) 252-5298, or e-mail
special.markets@perseusbooks.com.

1 2 3 4 5 6 7 8 9—08 07 06

*To my wife and girls—*
*Cindy, Amanda, and Tamara—*
*as always, for your love and support*

*And to the men and women of Project Delta,*
*who motivated me to pursue*
*directed energy so many years ago*

# Contents

# Acknowledgments

Directed energy (DE) encompasses a wide, cross-disciplinary field of science and engineering. It is nearly impossible to enumerate the many academic and technical disciplines that make up DE, as it includes fields as diverse as physics and engineering to psychology (for studying the Active Denial effect). The people who have advanced the research and development of DE are just as numerous, and it is not the intent of this book to acknowledge everyone involved; instead, I have chosen to highlight the efforts of a few individuals and hold them up as the DE "everyman." As such, I may have inadvertently forgotten to thank or mention everyone responsible for contributing to this book and the field of directed energy.

Special thanks go to Dr. Bill Baker, Dr. Kirk Hackett, Dr. Earl Good, Dr. Barry Hogge, Dr. Diana Loree, Lieutenant Colonel Chuck Beason (USAF, ret.), Stephanie Miller, Major General Don Lamberson, Ph.D. (USAF, ret.), Colonel Dick Tebay (USAF, ret.), Dr. Joe Miller, Commander George Bates (USN, ret.), CAPT Roger McGinnis, Ph.D., Dr. Bob Duffner, Dr. Barron Oder, John Albertine, Ed Pogue, Dr. Dinh Nguyen, Colonel Don Washburn, Ph.D. (USAF, ret.), Ray Saunders, Northrop

Grumman Corporation, TRW, Avco Space Systems, MDA, AirBorne Laser System Program Office, Lieutenant Colonel Brian Jonasen, Advanced Tactical Laser Program Office, Dr. Al Kehs, Colonel Ed Duff (USAF, ret.), Dr. Bruce Simpson, Cindy Beason, and Colonel Rich Garcia (USAF, ret.) from AFRL/DE Public Affairs for reviewing the manuscript. I also want to thank my agent, Matt Bialer, for his support in selling the book and encouraging me, my patient project editor, Erica Lawrence, and my editor at Da Capo, Robert Pigeon, for believing in this book.

DE research and development has been shrouded in a veil of secrecy. There are national security reasons for not revealing certain applications or vulnerabilities. My reviewers and I have been careful to ensure that no classified or insider information has been disclosed. I relied on publicly released information as well as interviews to build the story of directed energy. But even with these caveats, this book should serve layman and specialist alike with a broad overview of the people, the programs, and the products of what may prove to be the next "world-changing, disruptive technology."

Except where referenced, the views expressed in this book are solely those of the author and do not in any way reflect those of the Department of Defense, the U.S. Air Force, the Department of Energy, Los Alamos National Laboratory, or the Regents of the University of California.

—DOUG BEASON
Albuquerque
June 2005

# Directed Energy Milestones

## Prelaser

212 B.C.  Greek historian Lucian relates how a Greek general named Hippocrates, using an idea attributed to Archimedes of focusing sunlight with mirrors, sets fire to the sails of the Roman fleet at the siege of Syracuse

1870  John Tyndall demonstrates light refracting in a stream of water (early fiber optic)

1880  William Wheeling patents "piping light"

1880  Alexander Bell invents optical voice transmission

1898  Martians use directed energy (DE) death rays in *The War of the Worlds*, by H. G. Wells

1917  Albert Einstein discovers stimulated emission, the basis behind lasers

1945  Japanese-designed magnetron developed for possible HPM weapon[1]

1953  H. Motz and collaborators obtain coherent radiation from stimulated emission using a radio frequency (RF) linear accelerator to drive an electron beam through a series of alternating magnets

1956  O'Brian and Kapany (who invented the term "fiber optic") demonstrate all-glass optical fibers

1957  As a student at Columbia University, Gould describes the laser as an intense light source

1957    Townes, Gordon, and Zeiger invent the maser (microwave laser)

1958    Schawlow and Townes publish the first paper on the laser

## Postlaser

1960    Maiman builds the world's first laser (a ruby laser)

1961    Hall invents the semiconductor laser (precursor to the diode laser)

1963    First laser surgery performed at Stanford using a low-power argon laser

1964    Patel invents the molecular gas carbon dioxide ($CO_2$) laser

1965    Pimental invents the first chemical laser

1966    Gerry demonstrates the gas dynamic laser (GDL) at Avco Everett Research Lab, generating 138 KW by 1968

1968    ARPA (Advanced Research Projects Agency, predecessor to today's DARPA) establishes Eighth Card, a classified laser program to reach 500 kilowatts

1970    Mauer and his team invent the first practical optical fiber

1971    Hughes and Air Force Weapons Lab hit an aircraft with a $CO_2$ laser

1973    The air force's classified Project Delta downs an aerial drone with a high-energy laser

1975    IBM markets the first laser printer

1976    Army shoots down a drone and a helicopter at Redstone arsenal with an Avco electric discharge laser (EDL)

1976    Bell Labs demonstrates first multimode fiber optic

1977    McDermott and his team invent the COIL (chemical oxygen-iodine laser)

1978    Navy shoots down an army TOW missile with a TRW 400-kilowatt deuterium fluoride (DF) laser at San Juan Capistrano

1983    Airborne Laser Lab downs an AIM-9B Sidewinder missile with a $CO_2$ laser

1991    Artificial Beacon (adaptive optics) research at the Air Force Weapons Lab declassified

1993    USAF establishes the Airborne Laser System Program Office at the Phillips Lab

2001    Defense Department declassifies Active Denial, the first nonlethal DE antipersonnel weapon

2002    Army's THEL laser destroys Katyshu rockets, artillery rounds, and mortar shells at White Sands Missile Range

2003    ZEUS, the Army's anti–land mine laser, deployed to Afghanistan

2006    NIRF, the navy's anti-IED (improvised explosive device) high-power microwave, scheduled for deployment to Iraq

200?    First nonlethal millimeter wave Active Denial system scheduled for deployment

200?    Airborne laser scheduled to shoot down ballistic missile

As we enter the twenty-first century we find ourselves on the verge of a new breakthrough in warfare with the application of Directed Energy technology to the battlefield. As advanced sensors and kill mechanisms, Directed Energy applications in the laser and high power microwave areas will become the centerpiece of twenty-first century arsenals.

We are in an era in which precision and the lack of collateral damage are determinants in the acceptability of weapons. Directed Energy weapons with their ability to generate both lethal and non-lethal effects at the speed of light will gain greater acceptance.

The nation with the vision to embrace these weapons will dominate the battlefield for the foreseeable future. When combined with near real time intelligence, surveillance, and reconnaissance assets, the ability to strike quickly with . . . Directed Energy weapons will revolutionize warfare for surface forces. For an aggressor, sanctuaries will be few and retribution swift when faced with such revolutionary systems. . . .

—DUFFNER, BUTTS, BEASON, AND FOGLEMAN,
"Directed Energy: The Wave of the Future,"
in *The Limitless Sky: Air Force Science and
Technology Contributions to the Nation*

# A World-Changing,
# Disruptive Technology

THE DATE IS LATE FALL, and a cacophony of sound reverberates through the city—cars honking, animals braying, police whistles blowing. The air is dense, humid, and heavy with dung, car fumes, and urine. Beggars crowd the street, fighting for rupees given in embarrassed sorrow by widows, visiting dignitaries, and once enchanted students who now stare agape at the world's worst example of poverty.

The place is New Delhi, India, home of the world's largest democracy and unwavering friend to the United States.

Until now.

An unruly crowd surges through the trash-laden streets, picking up stragglers as the mob grows in frenzy.

Women and children slip around corners and cover their faces, trying to be unnoticed, but they are swept along with the roiling crowd. Unturbaned Indians stream into the street, dressed in garments that once would have been rejected by Goodwill, but now are the only clothes they own and, because of the incredible poverty, may be the only clothes they'll ever have.

Shouts erupt, rocks are thrown. Within minutes, the growing riot approaches the iron gates of the U.S. embassy.

A glass bottle filled with gasoline and stuffed with a burning rag is hurled over the gate. Burning liquid from the Molotov cocktail splatters across the ground. Someone shoots a gun and there is a panic. An unpredictable mob mentality forms as the crowd surges forward.

Stoic Marines guarding the entry points fall back into position, drawing their automatic weapons. After the debacle of the Iranian hostage situation twenty years before, the Marines are under unwavering orders to not give up the embassy, no matter what. Their orders are "shoot to kill." Their actions could set international relations back fifty years.

Hundreds of American and Indian staff members are hastily ushered into basement safe areas. The situation is escalating out of control. Some women and children in the crowd are jostled and thrown to the ground. The rioters are using others as human shields to prevent the Americans from stopping their advance. They know the Americans won't kill innocent women and children—like the Vietnamese who once stored weapons in hospitals to escape American bombing or Saddam Hussein, who placed military command and control centers in mosques. The rioting insurgents boldly push their innocent shields in front of them as they advance.

The crowd surges forward. The Marines must act.

The political balance with one of America's greatest democratic allies now hinges on the split-second decisions made by the gun-toting Marines, nineteen- and twenty-year-old men—just boys, not diplomats.

Visions of their fellow Marines being overrun at Fallujah and Mogadishu race through their heads. After the debacle of the U.S. embassy being overtaken by terrorists masquerading as students in Iran, the U.S. government swore that never again would an American embassy be overrun.

Never.

Young American soldiers, some still teenagers, are faced with immense pressure to react, with a consequence of killing innocent women and children.

Until now, the Marines only had two options: to shout at the insurgents, pleading with them to stop—or to shoot them. A simple binary decision. Shout or shoot.

America still reverberated from the decisions made nearly 30 years before, during the fall of the embassy in Iran.

With their orders in mind and no other option at hand, their course was ordained.

As the Marines raise their rifles, a deep humming sound envelopes the compound.

Suddenly the rioters feel intense heat, as if a gigantic oven had suddenly opened in front of them. Within seconds the pain is unbearable. They cannot think, they cannot reason—they can only react.

They turn and flee from the unseen heat source. Screaming in pain, the rioters drop their weapons as they sprint away. None look back as they scramble to leave the area.

Curiously, none of the women or children in the mob are affected.

As if divided by an all-knowing Solomon who can distinguish between hostile intent and innocence, only those people carrying weapons had felt the intense, excruciating pain.

In less than a minute the streets are clear and the compound is eerily quiet. The women and children disperse, unharmed.

The only noise in the U.S. embassy is the low mechanical thrumming that comes from a geodesic sphere inconspicuously located on top of the sprawling building.

Inside the sphere is the phased-array dipole antenna that directed the millimeter waves from the world's first nonlethal directed energy weapon, Active Denial.

......................

Science fiction? No. Active Denial is being tested today. Although this scenario didn't happen, it might.

And if funding had not been cut in the late 1990s, Active Denial could have been used to quell the urban warfare in Baghdad, Fallujah, and other cities in Iraq.

And countless lives might have been saved.

The size of the army matters, but technology wins wars.

At the height of the Roman Empire, Roman legions armed with arrows, long staffs, and shields used precise, steadfast formations to devastate the more numerous but ill-equipped barbarian hordes.

The invention of the stirrup in the sixth century gave a horseman the ability to use his mount as a lethal weapon—an astonishing transformation from the centuries-old use of transportation or plowing, allowing warriors to combine their horse's mass and speed with their devastating thrust of a spear.

In 1232, during the battle of Kai-Keng, the Chinese repelled Mongol invaders with the first known use of rudimentary rockets powered by gunpowder, called "arrows of flying fire."

In 1914, employing tanks and automatic weapons, the German army rolled across Europe, dominating the cavalry and breech-loaded armies that made up the conventional forces of that era.

On August 9, 1945, a lone American B-29 bomber flew over Nagasaki, Japan, and dropped a single atomic bomb that effectively ended World War II.

And in February 1991, precision-guided "smart" bombs, ground-hugging cruise missiles, and invisible stealth fighters forced the massively equipped and much more numerous Iraqi army to its knees.

In 2003, the conflict in Iraq just missed seeing the introduction of a new generation of sophisticated weaponry—the subject of this book.

These overwhelming victories had one thing in common—they exploited technology to incur a revolution in military affairs (RMA). An RMA is a revolution in warfare so dramatic, so disruptive, and so profound that it changes the way wars are fought, how nations interact, the way they conduct business, and even what drives their economy.

After their stunning victories, the Romans declared Pax Romana on the lands they conquered.

After World War II, the Cold War and the threat of nuclear winter drove the investment of billions of dollars into the defense–industrial complex and ignited hyperactive political movements.

And after the first Gulf War, terrorism grew and flourished, Saddam Hussein murdered thousands of Kurds, and al-Qaeda took down the World Trade Center—causing American policy to lurch away from the (once radical) Monroe doctrine to the (now radical) Bush doctrine of preemptive strike.

Today the next revolution in military affairs is about to begin.

But this revolution is not built on bombs or bullets, or anything you can hold in your hands. It's made of ordinary light—in the same spectrum of energy found in a microwave, a lightbulb, or a TV remote control.

It's called directed energy.

....................

The date is midsummer.

The place is Osan Air Force Base, home of 7th Air Force and the 51st Fighter Wing, located just 48 miles south of the demilitarized zone (DMZ). Negotiations have broken down and tensions between North and South Korea have never been higher.

The International Atomic Energy Agency has been barred from inspecting North Korean nuclear power plants, and although North Korea has claimed for years that it has nuclear weapons, only now is there indisputable evidence that enough plutonium has been siphoned away from the residual commercial fuel to construct several atomic bombs.

Forty-five thousand American troops are stationed on the 55-year-old DMZ along the 38th parallel. They are on highest alert as 500,000 South Korean soldiers back them up.

But facing them across the border are over 1.4 million North Korean regulars armed with an unknown number of Taepodong-3 ballistic missiles, in all probability tipped with nuclear warheads. And they can reach the western United States within forty-five minutes of launch.

Home to the last oppressive, totalitarian government in the world, North Korea reveals little about its capabilities or motivations. All that is certain is that the world is teetering on the brink of war.

Suddenly seven sleek missiles roar from silos deep in the valleys of North Korea. Three rockets streak to the south, arrowing toward Seoul and its 5 million inhabitants. The other four missiles veer east—heading for San Francisco, Los Angeles, Seattle, and San Diego.

Within seconds the missiles break through the cloud layer. In another two minutes they will exhaust the fuel in their upper stages and will soar unfettered to their targets in an arcing, parabolic trajectory. Officials estimate that 10–50 million deaths will occur over the next few days. Nothing can be done. The situation is hopeless.

Until now.

Orbiting at 40,000 feet above ground in a "racetrack" pattern 100 kilometers south of the DMZ, two specially modified 747s fly safely away from enemy fire. Infrared seekers on board the 747s pick up the bright rocket plumes as the Taepodong-3 missiles break through the cloud layer.

In milliseconds (thousandths of a second) low-power targeting and tracking lasers lock on to the missiles. Onboard computers calculate trajectories, and in the nose of the converted 747s, concave mirrors five feet across swing toward the rising missiles.

Roaring to life, giant turbopumps at the rear of the planes blast a supersonic mixture of hydrogen peroxide and iodine through specially designed nozzles, ready to power the world's largest laser. At the front of the plane, deformable mirrors shaped by hundreds of actuators embedded behind the mirror's highly polished surface change the mirror's surface hundreds of times a second. This is adaptive optics, invented by the military and now used by every major astronomical telescope in the world. Adaptive optics makes a perfect laser beam as the deformities in the atmosphere are taken out of the laser, even before the beam leaves the plane.

Thirty seconds after the Taepodong-3 missiles break the cloud layer, laser powers in the million-watt class of invisible energy streak from each of the 747s at the speed of light. The planes hold their beams razor sharp against the missiles, heating the metal skin with enough power to allow the missile's fuel tanks to explode from internal pressure within seconds.

One by one, infrared beams from the two airborne lasers jump from missile to missile, like picking off skeet targets at a shooting range. Exploding debris falls on enemy territory, leaving South Korea and the United States unharmed.

* * * * * * * * * * * * * * * * * * *

Again, is this science fiction? No. The first 747 airborne laser is undergoing flight tests today.

# 1

## The Promise of Directed Energy

---

DIRECTED ENERGY (DE) WEAPONS—lasers, high-power microwaves (HPMs), and particle beams—have come of age. Over the past two decades, directed energy power has increased by nine orders of magnitude—over a billion times—from milliwatt to megawatt. This is like supercharging a laser pointer used for highlighting PowerPoint slides to shooting down ballistic missiles 100 kilometers away.

Directed energy is making world-changing, revolutionary advances from fighting wars to battling terrorism. And it's doing so today. It's happening so fast that it's the equivalent of a military "future shock."

The first DE weapons are being developed, and in the next few years, when they are unleashed on the battlefield, they'll be more revolutionary than the longbow, machine gun, stealth airplane, cruise missile, nuclear submarine, or atomic bomb. The second Iraq War may well be the last *not* to depend on directed energy.

National leaders will soon have the ability to instantly deter threats anywhere in the world with infinite precision at the speed of light. The dynamic changes this will make to international relations will reverberate throughout American society. It will transform our way of life.

This is because directed energy is more than a new weapon in the warrior's arsenal. It's about a completely new way of thinking, a new way of employing both strategic and nonlethal force, and interacting in the international community.

Our large, mechanistic defense establishment, which served so well throughout the Cold War, will be transformed into a lighter, more agile, and information-centered force, shifting hundreds of thousands of people and billions of dollars from the government to the commercial marketplace. Over the next decade, the shift will result in the most profound change to the Defense Department since World War II. Just as tourism was revolutionized by the jet engine and communication was forever changed by the transistor, the next societal change will be fueled by directed energy, specifically directed energy weapons (DEW).

But does everyone share this view? And if directed energy weapons are so revolutionary, then why aren't they being championed as "the next big thing"? On the contrary, directed energy weapons have many critics; for example, the APS (American Physical Society, the world's premier organization of physicists) is skeptical of the benefits and capabilities of DEW and has sponsored several politically charged studies of the subject.

A major APS study was conducted in 1986 in response to President Reagan's Strategic Defense Initiative (Star Wars); the latest was in the fall of 2002 on America's ballistic missile defense, the Boost-Phase Intercept Study. This kind of criticism is not limited to strategic uses of laser weapons; high-power microwaves have their skeptics as well. Human rights advocates are up in arms about the long-term, unknown effects of Active Denial (the world's first nonlethal directed energy weapon) and the possibility of people on the ground receiving eye damage from the airborne laser as laser light glints off ballistic missiles when they are being destroyed.

Other questions swirl around directed energy weapons as they make their way to the battlefield: What happens if they proliferate? Someday other nations will surely obtain the technology; proliferation has always happened.

Are there any long-term effects that might occur to those exposed to DE? The memory of soldiers marching and flying into atomic fallout clouds, unsuspecting LSD and biowarfare test subjects, and other "safe" experiments burn brightly in the public's memory. Apart from its technical promise, directed energy's future is clouded by political and societal uncertainty. Will politicians ever allow it to be used under fear of possible unknown long-term effects?

....................

Deep in the arid desert, two Russian-made Katyusha rockets filled with enough high explosives to obliterate a building are launched toward friendly forces. The rockets are incredibly precise and can rain destruction from over 20 kilometers away. Israel considers them its number one conventional threat.

As the Katyushas soar toward their target, infrared trackers on the ground pinpoint their location. The two rockets' trajectories are quickly computed and fed to a battle-management system that steers laser beam directors to the target. A DF—deuterium fluoride—laser developed for the U.S. army hurls hundreds of thousands of watts of laser energy toward the rockets. Within seconds the Katyushas explode in midair, their debris falling harmlessly to the ground.

....................

The date for this scenario? Summer 2001—months before the September 11 terrorist attack on the World Trade Center.

The place? White Sands, New Mexico, only miles from Trinity site, birthplace of the world's first atomic blast, another revolution in military affairs.

To date, over 30 Katyusha rockets have been shot down in realistic scenarios such as this by THEL—tactical high-energy laser, the world's first high-energy laser weapon developed by the U.S. army and funded by the Israeli government for deployment along Israel's borders.

In November 2002 at the same White Sands range, THEL expanded its capability by shooting down artillery shells—something even most science fiction writers never thought possible. Recently even mortars have been destroyed by the THEL laser, resulting in the development of a mobile version of THEL (called MTHEL for Mobile THEL).

Directed energy is not science fiction. These are real weapons being tested in real scenarios.

Directed energy is maturing on a daily basis, and advances in technology are accelerating its use. The only reason these major DEW systems were not used in the second Iraq War was because they were still being tested and were not considered ready for the battlefield.

Largely shrouded in a highly classified environment, directed energy weapons research is conducted by a cadre of closed-mouthed technical wizards. The government labs that created revolutions in military weaponry in the past—nuclear weapons, stealth airplanes, and precision-guided weapons—have turned their talent to what they hope is their next ace in the hole—directed energy weapons. And they're on a path to move them to the battlefield. They're betting that before the world knows it, DEW will break into the headlines as it provides an overwhelming asymmetrical advantage in war.

And those nations that are not prepared to exploit directed energy will stagnate or, even worse, lose, by clinging to outmoded traditional forms of warfare. They will fall behind, just as civilizations that clung to the bow and arrow lost to the rifle and just as bullets and bombs will fall to DEW.

What about the future? We've already seen how lasers will be used in a spectrum of engagements, from shooting down ballistic missiles to protecting troops from mortar and artillery attacks. Will the military be satisfied with just fielding Active Denial on the ground, protecting embassies and airfields?

......................

Fast-forward to a decade from now. An American military plane flies low over a poor African village, dropping food, blankets, and medicine. Famine is killing thousands as the corrupt national government diverts Red Cross and U.N.-sponsored food deliveries to the black market.

As the low-flying plane makes a final delivery, a shoulder-launched surface-to-air missile spirals out of the jungle and hits the plane, exploding it.

Spiraling out of control, the plane crashes and the injured crew members are quickly captured. They are surrounded by jubilant troops, all loyal to the corrupt government.

Newscasters relive Captain Scott O'Grady's shoot-down over Bosnia and the disastrous "Black Hawk down" scenario as a live CNN feed shows bloodied American airmen being led blindfolded through a hostile village. Curious women and children surround the captive airmen and watch as they are beaten.

Sending in special operations forces is not an option. Although they performed brilliantly in Afghanistan, routing al-Qaeda from treacherous caves, there are too many innocent people surrounding the captives for elite Delta or SEAL teams to attack.

Things appear hopeless. A horrific massacre seems inevitable.

Suddenly the sound of a lone helicopter comes from high overhead as a special operations MH-53J helicopter approaches. A strangely shaped antenna protrudes from the helicopter's metal skin.

Although the helicopter is 10 kilometers away, out of range of shoulder-launched missiles, selected people in the village feel an intense sensation of heat coming from above, as if a supercharged oven had opened over their heads. The beam is precise, laserlike as it targets anyone approaching the captured crew. The aggressors are herded away from the hostages like sheep prodded by a stick.

People run, scrambling to get away from the incredible heat, which miraculously disappears once they are out of the village. Stepping back into the village perimeter instantly brings another hot, blasting shock. But once exposed to the heat, no one dares return.

Forewarned that Active Denial might be used in a rescue attempt, the entire flight crew hunkers down on the ground and waits for the rescue ship that is on the way. The ADS beam is extremely accurate, even at these distances. They experienced this situation before, in tests conducted at Kirtland AFB in New Mexico in their rescue training.

The noise around them suddenly stops, and the airmen lift their heads and look cautiously around. The village is completely clear, vacant.

Anyone who tries to approach the downed airmen is blasted with incredible heat from Active Denial and doesn't try again.

Within minutes, two other rescue helicopters swoop down into the village and pick up the stranded airmen as the lone MH-53 helicopter carrying Active Denial serves as a guardian, flying above, away from SAMs and antiaircraft fire.

......................

This scenario, along with many others, is part of the military's strategy for using the asymmetric capability of directed energy. But to accomplish this plan, investments in science and technology are needed to shrink the size of present-day Active Denial systems.

There are other possibilities for this scenario. Consider the same location, but instead of 2015, suppose the date is today.

......................

As the low-flying U.S. plane delivers its supplies, a terrorist's shoulder-held missile comes spiraling out of the jungle.

A heat sensor on board the American plane directs a solid-state infrared laser toward the oncoming missile. Within fractions of a second, the laser locks on to the warhead and starts jamming the missile's seeker. In seconds, the missile loses "lock" and veers away, out of control.

And no, it's not science fiction.

Although technically not a weapon, LAIRCM (large aircraft infrared countermeasure) is an antimissile laser jammer the USAF is installing on its aircraft today.

LAIRCM uses a relatively low-power laser not as a weapon but as a defensive shield. The key part of making LAIRCM work is exploiting advances in directed energy and jamming the target in a focused manner—all at the speed of light.

......................

When the laser was invented on July 6, 1960, everyone from military strategists to science fiction writers predicted that DE would be used as a weapon. But people were disappointed when lasers didn't cause a Buck Rogers blow-it-up effect. Tests showed that the most sophisticated lasers in the early 1960s only produced a low-power, although intensely brilliant, point of light.

The reason was that the technology for producing the laser was relatively immature. In the early 1960s, laser power levels were measured in thousandths of a watt. Typical laser pointers today, available for a few dollars at any office store, produce unwavering but low-power beams on the order of 5 milliwatts (or 5 thousandths of a watt), a hundred thousand times less powerful than the lightbulb shining in your hallway.

The difference between a lightbulb and a laser is that the incandescent or fluorescent bulb's energy is dissipated throughout the room. The bulb's light is projected homogeneously into every nook and cranny, as the photons—elementary constituents of all light—propagate, diffract, and scatter equally throughout the room, randomly traveling in every direction from the bulb.

It is just as likely for a photon to be emitted toward the front of the room as toward the back or the floor. Figure 1.1 shows a simplified diagram of a homogeneous and isotropic (equivalent in all angular directions) emitting lightbulb.

On the other hand, the photons in a laser all move in the same direction, exactly parallel to one another, in lockstep and with no angular deviation. This is a result of the quantum response of excited atomic and molecular states.

Figure 1.1    Lightbulb emitting photons equally in all directions—a random, noncoherent, non-phased source of light.

Even more remarkable is the fact that these photons are also traveling with the same phase. They oscillate in time with one another, much the same way as armies march in step throughout an entire formation.

DE first proved useful in low-power applications, allowing for the invention of microwave ovens, CD players, TV remote controls, fiber optics, DVD technology, and laser eye surgery; the list goes on. As DE power increased, the applications for this technology grew exponentially. Today DE is used in heavy industry for welding metal, cutting steel, and drilling through rock, doing these tasks more cheaply, quickly, and safely than traditional methods.

For the military–defense complex, DE heralds even greater promise. The reason is that bullets and bombs have reached the limit of their capability. In World War II approximately 5,000 bombs were required to destroy one target.[1] In Vietnam, with the advent of laser-guided technology, that number dropped to around 500. Precision-aiming technology advanced, and by 1991 in the first Iraq War it took approximately 15 bombs to destroy a target; in Kosovo and then Afghanistan, that number dropped from 10 to 5 bombs. Even more precise weapons were used in the 2003 conflict with Iraq, and ratios approached one target killed for every weapon dispensed.

However, with the ultimate limit of one bomb being used to destroy one target, warriors can't do any better. They will still be limited by the number of bombs they can carry, even though a single weapons system such as the B-2 can hit dozens of targets per flight.

Another drawback is that bombs and bullets reach their target by following the law of gravity. They travel in trajectories constrained by ballistics and thus take a finite time, sometimes up to minutes, to reach their target.

This is where directed energy weapons can radically change the nature of warfare, and why national and military leaders are so excited about their use: they can be applied to a target almost instantly, thousands of times faster than any conventional weapon.

Directed energy travels at the speed of light, 186,000 miles a second. This velocity is incomprehensible to anyone who is used to the normal world where people jog at 3 miles an hour, cars zip down the highway at 65 miles an hour, and the fastest airliners traverse the Atlantic at speeds approaching 700 miles an hour. The world's absolute speed record is held by astronaut Lieutenant General Tom Stafford, commander of *Apollo X*. When his spacecraft returned from orbiting the moon, it achieved 28,547 miles per hour, around Mach 38, or 8 miles a second—0.002 percent of the speed of light, still the world's all-time speed record for a human.[2]

Light, whether produced by the sun or in a lightbulb, travels fast enough to circle the earth over seven times in a second, traveling at approximately Mach 1 million. That means that directed energy—light that is in the form of lasers or microwaves—can reach its target almost instantly.

In military terms, a weapon's effectiveness is measured by the equivalent muzzle velocity. We'll see in Chapter 3 a better way to measure military effectiveness, but comparing velocities is a good start. A bullet's muzzle velocity may be as high as 6,000 feet per second, and thus DE's "muzzle velocity" is greater than 982 million feet per second—over 160,000 times faster than a typical bullet.

To grasp the implication of this speed, consider an enemy jet traveling 1,000 kilometers an hour (621 miles an hour). If you launch a conventional air-to-air missile traveling 2,000 kilometers an hour at the jet from 300 kilometers away (about 186 miles), it will take six minutes for that missile to reach the enemy jet. During that time—which is an incredibly long time for a fighter pilot—the fighter jet can make numerous evasive maneuvers.

However, instead of using an air-to-air missile, suppose you blast the jet with a laser weapon. Instead of minutes, it takes just 1/1,000 of a second for the beam to reach the enemy jet 300 kilometers away. In that length of time the jet would have traveled a third of a meter, or about a foot, leaving no time for an evasive maneuver. This allows today's warriors to "reach out and touch someone" instantly, and as we will see later in the book, even around the globe.

Another advantage to DE is that it can "flood" areas, allowing one DE weapon to defeat hundreds or even thousands of targets, as opposed to the absolute limit of one bomb killing one target. This gives the military the ability to carry a "deep magazine" and thus shorten the so-called logistics tail of ferrying a crate of bullets or bombs from the factory to the war zone to the fighter—and still having a maximum of one bomb killing one target.

World-changing events are fueled by revolutions in military affairs, and they are brought about by the invention of disruptive technologies so profound that they forever change the nature of society. Directed energy is the impetus for the next revolution, and it will change strategy and national policy, and ultimately affect billions of dollars in funding for the military services and homeland defense. In this book I will demonstrate how disruptive technology increases military effectiveness and how national policy and the disposition of billions of dollars have already been influenced by the emergence of directed energy.

In 2004 and 2005, the number one priority of the tactical military commander in Iraq was stopping deaths caused by IEDs (improvised explosive devices). These homemade bombs were disguised and hidden along the road where they were remotely detonated by insurgent terrorists.

Directed energy played a role in neutralizing these weapons, with the proposed battlefield introduction of the navy's NIRF (neutralizing IEDs with RF) HPM weapon. Along with the NIRF, the army's ZEUS—a powerful laser mounted on a combat vehicle—gave U.S. forces the ability to remotely detonate land mines.

Another priority in Iraq was to deploy a working Active Denial system (ADS) to the battlefield. Although the ADS has not to date been deployed, everyone agrees that its presence in Baghdad might have saved dozens of lives. Ironically, only a few years before, the program was nearly killed by jealous bureaucrats and dismissed by war fighters as irrelevant.

The U.S. Marines have said Active Denial is their most important nonlethal weapons program. If they had possessed it in Mogadishu, it would have saved American lives. And if we had it today, it might change the nature of urban warfare in Iraq.

Another example of the need for directed energy weapons comes from Iraq. Scud missiles are relatively short-range weapons of terror, on the order of several hundred kilometers, and are so inaccurate that they can only be launched against cities, not smaller targets such as military installations, ammunition warehouses, or command-and-control facilities.

However, in the first Gulf War the fear created by incoming Scuds conjured up sickening memories of World War II London besieged by Nazi V-2s. And, like the V-2s, the Scuds sometimes caused more than psychological damage, such as destruction of the American barracks in Saudi Arabia.

Frightened by America's inability to respond to the Iraqi Scud threat, air force chief of staff General Ron Fogelman and air force secretary Sheila Widnall fought for the airborne laser—only to have low-ranking acquisition officials try to kill the laser system and divert the funding to other controversial programs, such as adding additional capability to the F-22 fighter. Clearly not everyone in the military supports directed energy programs.

As many people are supportive of DE as are wary of it. From the ardent scientist to the aroused protestor, there seems little middle ground. So why are people so passionate about directed energy? Is there a historical basis behind this technology, or is it something entirely new?

Part of the answer lies in assessing how directed energy increases military effectiveness; if DE is nothing more than vaporware, a sophisticated light show that can dazzle but not destroy the enemy, then its military effectiveness is low and it is a waste of money. It's easy to promise the world and later have an excuse why the technology won't work. But on the other hand, if it provides the advantage that can help win wars, then it is worth supporting and getting excited about.

The application of stealth to a weapon platform is an example of a militarily effective technology. Stealth fighters led the campaign against Baghdad in the first Gulf War, and in concert with high-precision weapons they successfully negated the air defense of the most heavily defended city in the world.

But before we consider the military effectiveness of directed energy, we need to look at exactly what it is. It sounds mysterious, even incomprehensible, but in the next chapter I will supply an explanation—and it is simpler than you may think.

# 2

# What Exactly Is Directed Energy?

DIRECTED ENERGY SOUNDS EXOTIC, complicated, and mysterious, and therefore to many seems incomprehensible. After all, if this is a new phenomenon, then it must be complex, especially if it has to do with emerging science.

Consider an analogy from another field. In the prelude to this book, I noted that in 1232 during the battle of Kai-Keng, the Chinese repelled Mongol invaders with the first known use of rudimentary rockets powered by gunpowder, "arrows of flying fire." People were scared of this technology, mostly because they didn't understand it. We can reason that the Mongols were not as advanced as the Chinese and didn't comprehend the science. Therefore it would seem reasonable to expect, after nearly 800 years of being exposed to rocket technology—including the harrowing years of World War II with the German V-2s and the intercontinental ballistic missiles of the Cold War—that people would finally understand the basics of rocketry.

But even today rocketry produces an image of complexity. For example, although our national anthem explicitly refers to "the rockets' red glare," after all these years, most people find anything having to do with rockets totally incomprehensible.

Viewing rocketry with wariness and holding it at arm's length because of a perceived degree of difficulty has even crept into our modern language: how many difficult ideas are tagged with the caveat "it's rocket science"?

With directed energy, nothing could be further from the truth. In fact you encounter directed energy in many forms every day, even though you may not recognize it. Directed energy is present in the tiny laser in your CD and DVD player and in stoplights that use light-emitting diodes; in the laser light that courses through fiber optic telecommunications lines and even in the laser pointers available at any office supply store.

Microwaves are just as prevalent, from the common microwave oven to the microwave communication tower that transmits signals over large distances.

Given these examples, most people would grudgingly agree that directed energy is not really that mysterious, since it is used in everyday household items.

Yet to many there is still an incomprehensible difference between the sighting laser used in a hunting rifle compared to the high-power laser cannon that can knock a ballistic missile out of the sky. The former is familiar to hunters and to anyone who has watched a Hollywood action movie over the past decade; the latter seems to belong in the realm of science fiction.

But, aside from the obvious difference in power, is there really any difference between a sighting laser and a laser cannon?

Is the laser pointer you use in presenting the latest sales figures really related to the army's mobile tactical high-energy laser that has shot down rockets and artillery shells in White Sands?

The answer is that there is no difference between the two except for the power. Laser weapons and high-power microwaves are merely different aspects of directed energy—just more energetic than the common household applications listed above.

A laser weapon is defined as any laser used against an enemy. Relatively low-power lasers, not much more energetic than commercial laser pointers, may fall into this category if they are used to guide high-precision bombs. But generally a laser weapon belongs to a category of lasers with power levels ranging from 50 kilowatts to over a megawatt.

Tactical laser weapons, such as the Defense Department's ATL (advanced tactical laser), have a power level at the low end of the spectrum in the tens of kilowatts; strategic laser weapons, such as the Missile Defense Agency's ABL (airborne laser), is at the higher end and is known as a "megawatt-class" laser.

High-power microwave weapons are defined as "devices designed to disrupt, degrade or destroy targets by radiating electromagnetic energy in the RF [radio frequency] spectrum, typically between 10 MHz and 100 GHz."[1] The exact power level needed to produce these effects is classified, and the only hint to their magnitude is that they're called high-power microwaves instead of low-power microwave weapons.

## What Is Directed Energy Made Of?

What exactly makes up a laser? And for that matter, what is a microwave, besides being a tiny wave, as the name implies?

And going one step farther, how similar are lasers and microwaves? And if they are similar, why don't people talk about them as different facets of essentially the same thing? For example, although quarters and dimes are different, they are both coins—money—and therefore similar.

The answer is that in their simplest form, both lasers and microwaves are different manifestations of light—the same light that comes from the sun, is emitted by a lightbulb, or emanates from your TV.

Visible light is a type of radiation that we can see. In its most generalized form, light can be lumped together with microwaves, radio waves, X rays, gamma rays, and even heat (also known as infrared radiation), and is commonly classed as electromagnetic energy.

In 1856, James Clerk Maxwell wrote his groundbreaking paper show-ing the connection between the electric and magnetic fields. In this paper he proposed the existence of electromagnetic waves and theorized that they would travel in free space at the speed of light. Electromagnetic waves were actually measured 32 years later, when the German scientist Heinrich Hertz detected them in his laboratory.

The connection between electric and magnetic waves was later for-malized through what is known today as Maxwell's equations. These equations describe the propagation and interaction of what is now known as the electromagnetic (EM) spectrum—waves of EM energy that span phenomena from radio waves to X ray and gamma waves.

In the following sections I'll present some equations for readers who are interested in the mathematical details underlying directed energy. But don't let these equations scare you. If they make you uncomfortable, you can skip directly to the next chapter or you can skim over the equa-tions and simply read the text.

The main point is that lasers and HPM are essentially the same—they're just different parts of the electromagnetic spectrum. They're created somewhat differently and they have different characteristics, but the upshot is that they're essentially the same.

A traditional Maxwellian view of electricity and magnetism describes the EM spectrum through the relationship of different wavelengths of light. The length of the wave ($\lambda$) is measured in meters.

Think of the Greek symbol $\lambda$ as being a length, a distance. Mi-crowaves have $\lambda$ wavelengths of centimeters (1/100 of a meter) in length; laser wavelengths have $\lambda$ wavelengths 10,000 times smaller, or a micron (a millionth of a meter) in length.

Light from the sun is emitted in wavelengths ranging from hundreds of kilometers (which is in the long wavelength range transmitted by power lines) to cell phone wavelengths (in the millimeter range) to den-tal X ray wavelengths (a millionth of a meter) to gamma rays (a hundred millionth of a meter).

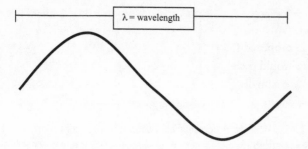

Figure 2.1    A generic wave of wavelength λ, where λ is a length, or distance, that may be several hundred kilometers long or as short as a billion billionth of a meter.

There is an inverse relation between wavelength λ and frequency υ (the number of times a wave crest passes through a certain point) given by

$$\lambda = c/\upsilon$$

where for electromagnetic waves, c is the speed of light, or $3 \times 10^8$ meters/second. Figure 2.1 shows a generic wave of wavelength λ.

Visible light is composed of a rainbowlike spectrum ranging from dark red, to orange, to yellow, to green, to blue, to purple, shown in Figure 2.2. This part of the spectrum covers wavelengths ranging from 700 nanometers (i.e., 700 billionths of a meter) to 400 nanometers (400 billionths of a meter).

These colors may also be expressed in terms of frequency derived from the inverse relationship of wavelength and frequency given above.

So why do scientists and engineers refer to light in terms of wavelength and frequency, rather than one or the other? Why on earth do they make things so difficult? After all, picking one or the other seems relatively straightforward, rather than flip-flopping between the two.[2]

Well, there's no easy answer except that some people are more comfortable using frequency units than wavelengths. Consequently, you'll hear some engineers talk about "millimeter waves." Others may talk about "terahertz waves." Still others may speak of "X band radiation."

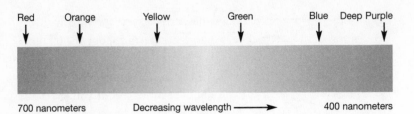

700 nanometers       Decreasing wavelength ⟶       400 nanometers

Figure 2.2  Visible light consists of a spectrum of colors, dependent on the light's wavelength (depicted here in grayscale). The longest wavelength (dark red, approximately 700 nanometers) is on the left; as the wavelength shortens, the colors change from red to orange, to yellow to green, then to blue and deep purple, the shortest wavelength (approximately 400 nanometers).

But they're all talking about high-frequency microwaves. Millimeter waves, terahertz radiation, and X band are all different ways of saying the same thing.

It's like getting a German, an Englishman, and an American together talking about euros, pounds, and dollars. They're all talking about money, but in different currencies. The information they exchange depends on where they came from and how comfortable they are in communicating across cultural lines.

In the same way, scientists and engineers confer in units with which they're comfortable. It would be nice if everyone spoke the same language, but this is just another example of why directed energy appears to be such a tough subject.

So in a way, we scientists and engineers do this to ourselves. It's not enough to make directed energy tough by wrapping it in equations and demonstrating its effects in a laboratory complete with white-coated technicians. No, we have to compartmentalize even the way we talk into a specialized language. But bear with me. It really isn't that tough.

I've touched on how lasers and microwaves are different manifestations of essentially the same thing—light, but at different frequencies (or, equivalently, wavelengths). Both lasers and microwaves are made up of electromagnetic radiation, but lasers are mostly visible or infrared light (400–700 nanometers), while microwaves are mostly high-frequency radio waves (millimeters to centimeters), about 10,000 times longer than lasers.

A more physical way to think of this is to take a long piece of string. Hold one end and have a friend hold the other end. If you shake the string, eventually you'll notice that a constant up-down-up-down wave pattern will be produced. These "standing waves" (called standing because they're not moving) have a certain length—the wavelength. Now shake the string faster. You're increasing the frequency. Because the frequency is increasing, the wavelength (or distance between crests) should grow smaller.

Call these high-frequency, short-wavelength waves a laser.

Now shake the string more slowly. Eventually a standing wave pattern will be built up. These waves are now longer than the last waves you produced.

Call these waves microwaves.

So what is the difference between the two, this "laser" and "microwave"?

Nothing, except for how fast you shook the string.

Same string, different wavelengths.

Lasers and microwaves. Same electromagnetic radiation, but with a difference in wavelength (or frequency).

Table 2.1 on the following page shows the domain of electromagnetic waves and the corresponding frequency and wavelength.

## A Quantum Interpretation of Light

Although electricity and magnetism can be described as waves, another equally correct interpretation is to describe them as tiny, quantized (discrete) packets of energy known as photons.

Lasers and high-power microwaves (HPM) are both made up of photons, and the only difference between the two is their energy (and thus their wavelength and frequency). A photon's energy (in units of joules) is a function of frequency $\upsilon$:

$$E = h\, \upsilon$$

Table 2.1    The Electromagnetic Spectrum

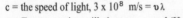

$c$ = the speed of light, $3 \times 10^8$ m/s = $\upsilon \lambda$
$\upsilon$ = Frequency, in oscillations per second (Hz)
$\lambda$ = Wavelength, in number of waves per meter (m)

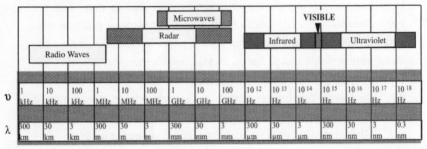

| | | | | Microwaves | | | | | | | VISIBLE | | | | |
| | | | Radar | | | | | | | Infrared | ▼ | | Ultraviolet | | |
| Radio Waves | | | | | | | | | | | | | | | |
| $\upsilon$ | 1 kHz | 10 kHz | 100 kHz | 1 MHz | 10 MHz | 100 MHz | 1 GHz | 10 GHz | 100 GHz | $10^{12}$ Hz | $10^{13}$ Hz | $10^{14}$ Hz | $10^{15}$ Hz | $10^{16}$ Hz | $10^{17}$ Hz | $10^{18}$ Hz |
| $\lambda$ | 300 km | 30 km | 3 km | 300 m | 30 m | 3 m | 300 mm | 30 mm | 3 mm | 300 μm | 30 μm | 3 μm | 300 nm | 30 nm | 3 nm | 0.3 nm |

km = kilometer ($10^3$m)   mm = millimeter ($10^{-3}$m)   μm = micrometer ($10^{-6}$m)   nm = nanometer ($10^{-9}$m)
KHZ = kilohertz ($10^3$Hz)   MHz = megahertz ($10^6$Hz)   GHz = gigahertz ($10^9$Hz)

where h is Plank's constant (h = 6.6238 x $10^{-34}$ joules-second) and $\upsilon$ is the frequency (or oscillations per second), and the wavelength and frequency $\upsilon$ are related by the equation

$$c = \lambda \upsilon$$

This is profound. I've been saying all along that lasers and microwaves are made up of the same "stuff"—electromagnetic radiation—and are similar except for their wavelength. If this stuff is really a photon, then the only difference between lasers and microwaves is their energy.

An equivalent way to say this is that since lasers and microwaves only differ in their energy, they must have different wavelengths. As it turns out, all the various characteristics of directed energy can be explained simply by this difference in wavelength.

Lasers and HPM are created in different ways (to obtain this difference in wavelength). Although this difference in wavelength may not seem like much, it causes lasers and HPM to travel through the atmosphere in different ways, to interact with matter in different ways, and to produce different effects on targets.

Part of this behavior rises from their dual nature of being waves as well as photons; but the majority of these dissimilarities are simply due to their difference in wavelength.

So there you have it. Directed energy is nothing more than different parts of the electromagnetic spectrum, where lasers have much shorter wavelengths than microwaves. In Chapter 5 we'll consider some additional differences between lasers and high-power microwaves; but first we'll see historical trends that indicate why directed energy may be that world-changing, disruptive technology I hypothesized in the Prelude.

Chapter 3 will present arguments why directed energy can be a militarily effective technology, as well as why there are major problems with directed energy—not unique to its technology but intrinsic to all new weapons that force war fighters to use new tactics and employ different doctrines. Chapter 4 will focus on specific drawbacks of directed energy unique to each technology.

# 3

# The Military
# Effectiveness of Directed Energy

WHY USE DIRECTED ENERGY (DE), and why spend millions of dollars fielding a weapons system that a few years ago was the staple of science fiction writers? In the mid-1980s, Senator Ted Kennedy publicly denounced the use of DE weapons for the Strategic Defense Initiative (SDI) known as Star Wars. And since then, directed energy weapons have met resistance from different quarters, such as the American Physical Society, which warned in 1987 that "even in the best of circumstances, a decade or more of intensive research would be required to provide the technical knowledge needed for an informed decision about the potential effectiveness and survivability of directed energy weapon systems."[1]

People are always suspicious of a new product or technology, especially when it is championed as solving their biggest problem: "A sure cure for baldness!" "Lose weight without dieting!" "Improve your sex life!"

The dot-com craze of the late 1990s hardened many to sweeping promises, especially when they were lacking in detail or were not backed up by a good track record. Software firms, for example, are still looking for the next "killer application," and investors have grown weary of empty promises.

So what makes directed energy any different? Lasers have been around since 1960 and microwaves have been used since World War II, but they haven't changed the world, except for radar, which was one of the technologies that helped win World War II. So why shouldn't we think that directed energy won't be another flash in the pan technology that the military–industrial complex is trying to shove down the tax-payer's throat?

In the first decade of the twenty-first century, with company profits down, the economy struggling to recover from a recession, and the echoes of the Cold War guiding national policy, why shouldn't we view directed energy as the defense equivalent of a dot-com, trying to attract investors with an alluring killer app?

The answer is that there is historical evidence for believing that directed energy will increase military effectiveness, and in the following discussion I will show how advances in science and technology can radically change the way the military fights war.[2] Later we'll see how investing in basic science has accelerated that trend, and why directed energy is the next logical step for increasing military effectiveness.

Since the beginning of World War II we have seen the introduction of precision-guided weapons, the atomic bomb, the ballistic missile, computers, jet aircraft, satellites, radar, cell phones, the global positioning system (GPS)—the list of scientific and technical applications in warfare is staggering.

And the pace of technological development is increasing. In the millennia since records have been kept, it is estimated that the world saw a doubling—100 percent growth—in knowledge from the dawn of time until the 1950s. And that knowledge has doubled several times since the 1950s.

Manpower Density on the Battlefield

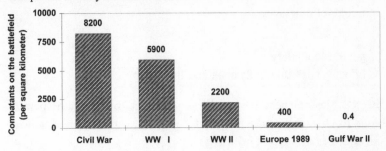

Figure 3.1    Manpower density on the battlefield (per square kilometer). Fewer people are needed to fight wars than in the past. This is primarily due to the increase in technology available to the war fighter. *Source:* Kenneth L. Adelman and Norman R. Augustine, *The Defense Revolution: Intelligent Downsizing of America's Military*; CNN for the second Gulf War[3]

That growth of knowledge has spilled over into war fighting. Today's warrior is fighting with more technologically sophisticated weapons, and consequently fewer warriors are needed on the battlefield.

But technological advances in warfare have been a double-edged sword, for while the number density (the number per square kilometer) of combatants may have decreased throughout the years, their firepower has increased.

Figure 3.1 shows the dramatic decrease in number density made possible by the introduction of new weapons. The increase in firepower may be understood by considering the way technology has enabled fewer war fighters to levy more damage at a longer distance: the range of a spear was extended by the bow and arrow; that range was extended by a bullet; that range was extended even farther by an artillery shell; and that range was extended by missile technology.

With directed energy weapons engaging at the speed of light, a weapon's range can be extended around the world, reducing manpower density on the battlefield even further. In Chapter 12, a real-world example is shown of a space-based relay mirror focusing laser energy hundreds of miles away.

## Increased Military
## Effectiveness Due to Science and Technology

In 1945 Major General J. F. C. Fuller enumerated five qualitative parameters to characterize the power of a weapon, with "range of action" identified as the highest priority:[4]

1. Range of action
2. Striking power
3. Accuracy of aim
4. Volume of fire
5. Portability

Brigadier General Simon P. Worden expanded on this concept by deriving military effectiveness as a basic measure of a weapon's military power.[5] Effectiveness may be defined in terms of the brightness (a term frequently used by laser engineers to measure the capability of a laser) per unit time, or the measure of a weapon's range, accuracy, and power per unit time, all rolled into a single number. Weapon effectiveness is listed in Table 3.1, and is illustrated in Figure 3.2.

Note that military effectiveness is presented in a compact form as an exponential number—meaning that bullets have a military effectiveness $10^2$, or 100 times greater than arrows; and that ICBMs are $10^8$, or 100,000,000 times more effective than artillery in 1900.

Worden projects lasers to be 10,000 million times more effective than artillery. Although military tactics and strategy have played a role in improving the effectiveness of these weapons, advances have chiefly been due to the exploitation of science and technology.

Figure 3.2 shows the dramatic increase in military effectiveness on a logarithmic scale; the y-axis is shown as exponential powers of 10, so

Table 3.1    Weapons Effectiveness[6]

| Era Year | Weapon | Time[a] | Brightness (joule/Sr) | Firing Rate (per sec) | Effectiveness[c] (joule/Sr/sec) |
|---|---|---|---|---|---|
| 1000 | Arrows | 6 Months | $10^8$ | $10^{-2}$ | $10^6$ |
| 1500 | Bullets | 3 Months | $10^9$ | $10^{-1}$ | $10^8$ |
| 1800 | Artillery | 1 Month | $10^{12}$ | $10^{-1}$ | $10^{11}$ |
| 1900 | Artillery | 1 Week | $10^{14}$ | 10 | $10^{14}$ |
| 1930 | Aircraft | 1 Day | $10^{19}$ | $10^{-1}$ | $10^{18}$ |
| 1950 | Aircraft | 1 Day | $10^{23}$ | $10^{-2}$ | $10^{21}$ |
| 1970 | ICBM | 1 Hour | $10^{23}$ | $10^{-1}$ | $10^{22}$ |
| 2015 | SBKKV[b] | 1 Hour | $10^{23}$ | 10 | $10^{23}$ |
| 2020 | Laser | 5 Minutes | $10^{22}$ | $10^2$ | $10^{24}$ |

Source: Simon P. Worden, SDI and the Alternatives.
[a] "Time" refers to both the time period of battle and the time it takes to get into position to engage the weapons.
[b] Space-based kinetic kill vehicle (SBKKV).
[c] Effectiveness = Brightness x firing rate.

that the maximum value of "25" is not a simple factor of 5 greater than "20," but rather $10^5$, or 100,000 times greater.

When will this stop? At the present rate, not for the foreseeable future, since the technology present on the battlefield keeps advancing.

Tomorrow's battlefield will consist of globally interconnected networks keeping track of targets using distributed, sophisticated, smart, and reconfigurable sensors; microcombatants; stealth air/land/sea and space platforms; and long-range, conventional (nonnuclear), high-precision, extremely accurate weapons systems (both manned and unmanned)—all linked with digital computers. And a key enabler to these systems will be directed energy because of its ability to deliver energy with nearly infinite precision, without regard to gravity or ballistic motion, all at the speed of light.

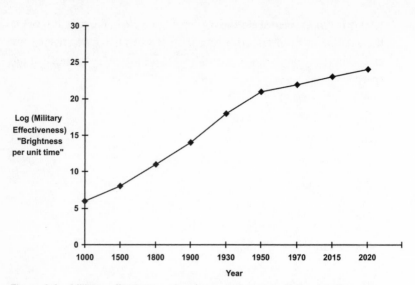

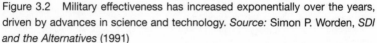

Figure 3.2    Military effectiveness has increased exponentially over the years, driven by advances in science and technology. *Source:* Simon P. Worden, *SDI and the Alternatives* (1991)

History has shown that advances in science and technology result in exponential increases of military effectiveness. Not just increases in 10 percent, or even a doubling of effectiveness. But true factors of many thousands of times.

The next step in this growth will be applications of directed energy, primarily in the form of lasers and high-power microwaves. Advances in science and technology will make their way to the battlefield and will change the nature of warfare. When lasers and high-power microwaves are introduced on the battlefield, they will be supplementing the weapons of the past—bullets, bombs, and missiles.

Just as when these older weapons were first introduced, directed energy won't entirely replace existing weapons; for example, today's soldiers are still using bayonets, although they may be armed with modern rifles outfitted with laser designators. And bullets are still fired by state-of-the-art jet fighters that carry sophisticated air-to-air missiles.

When it is employed on the battlefield, directed energy will provide the winning, asymmetric advantage to increase military effectiveness by extending a weapon's range, precision, and ultimately lethality, all at the speed of light.

# 4

## The Problem with Directed Energy

By NOW YOU MAY HAVE THE IDEA that directed energy can solve all the world's problems, like a technological equivalent of Superman: it's faster than a speeding bullet (travels at the speed of light) and can leap buildings with a single bound (is not constrained by gravity and can be reflected by mirrors in space). But just as the 1950s Superman image glosses over basic human limitations, the rosy outlook for directed energy glosses over its own problems. And some of them are what the entertainment field calls showstoppers.

All programs have drawbacks, and no new weapon program, especially in modern times, has ever been fielded without overcoming them.

The fledgling U.S. rocket program had a failure rate approaching 60 percent in its infancy, and programs now viewed as wildly successful, such as the advent of stealth technology, were fraught with problems.

For a revolutionary technology such as directed energy, which not only allows a new way of fighting but dictates changes in doctrine, tactics, and strategy, one must be steeled to expect major problems. Otherwise, Pollyannish expectations of success will be shattered and unrealistic demands will be placed on directed energy that are not being made on other major technological programs, such as the F-22.

There are three major problems with fielding directed energy weapons: transitioning DE to the battlefield, the economics of using DE versus traditional or conventional weapons, and the tendency of DE proponents to oversell DE capabilities. These create false expectations and give DE a bad reputation before its has a chance to prove itself.

## The Problem of Transitioning DE

The biggest problem facing directed energy is how to successfully transition science and technology out of the laboratory and into the hands of the war fighter.

This problem has many parts, the first being the need to mature the technology from a neat science experiment into an engineering reality that can be used outside of a perfectly-controlled laboratory environment and can be exploited by 19-year-old soldiers instead of geeky physicists.[1]

The laboratory environment is an idealized world in which pristine conditions of humidity, temperature, density, airflow, lighting, and even the size and number of dust particles can be strictly controlled. In these clinical conditions, scientists need not worry about how external factors affect the science experiment; rather, they focus on what the experiment is doing and why it is proceeding as it is.

The scientific method dictates that experimental results must be duplicated for a theory to be proven true, and the laboratory environment is idealized to allow the researcher to concentrate on the science, not the environment. And once a scientist has succeeded in duplicating the results, the experiment is deemed a success.

But generating kilowatts of laser power in a lab does not mean that the laser is ready for prime time, but only that building a laser weapon may be possible. An actual laser weapon must operate on the battlefield at a moment's notice, after being bounced in a jeep (or shoved inside a plane, or sloshed onboard a ship), fully fueled and ready to lase in darkness, in the rain, in a dust storm, or in 120-degree desert heat, the first time and every time so the war fighter can use it to win the war. Otherwise it's just a useless piece of junk. A science experiment.

It takes time and money to transition science out of the lab and build in the "ilities" needed by the war fighter—maintainability, refuelability, reliability, and so on.

Although the military has established processes to transition technology out of the laboratory and into the battlefield, more often than not, once demonstrated in a lab the science and technology are just thrown "over the transom" to a well-meaning but ill-equipped contractor who must reengineer the technology to make it work in a war fighting environment.[2]

The second problem with transitioning directed energy occurs once it's out of the laboratory—how do you get people to use it?

Not everyone on the battlefield will be thrilled to use DE, even with the exciting capability it promises. After all, war fighters choose weapons because they work, not because some lab rat told them the weapons would change the very nature of warfare. Ask a soldier if he'd rather have a howitzer or a strategic laser system, and there'd be no surprise in his answer.

A war fighter's weapon has proven itself—sometimes in a battlefield environment and sometimes in realistic simulations. Asking a soldier or an airman to exchange a bomb, a bullet, or a missile for a directed energy weapon—something the war fighter can't even see much less comprehend—takes a long leap of faith, a leap most war fighters are not willing to take.

War fighters don't always know what a new weapon can do or what it is capable of doing. (There is more on this later, in a section aptly named "Give a pilot a ball of crap and a string, and he'll make a game of it.") For example, war fighters have had several decades to learn that air-to-air missiles can be fired at aircraft while the enemy is miles away, even when the enemy can't be seen; or that sometimes it is better to fire a volley of bullets at the enemy when the enemy is too close, than to use an air-to-air missile.

Fighter pilots didn't discover these tactics right away—it took years for them to develop.

The air force Red Flag exercise in Nevada was established to train pilots in realistic situations, to teach them tactics and try out new techniques of using these weapons. The army, navy, and marines have similar training courses, all focused on allowing the war fighter time to experiment and become an expert in using a new weapon's capabilities.

To take a lesson from history, during World War I, the airplane was first used primarily as an observation platform—an extension of a manned observation balloon that could look out over an enemy's position. Airplanes flew over enemy positions because their main mission was surveillance. Pilots only later started carrying bombs, an increase in mission capability that was incomprehensible to most and dismissed as inconsequential by many.

For example, in 1921, when Secretary of War Newton Baker was told of Billy Mitchell's claim that airplanes could sink battleships, he growled, "That idea is so damned nonsensical and impossible that I'm willing to stand on the bridge of a battleship while that nitwit tries to hit it from the air."

In 1938 Major General John K. Herr said, "We must not be misled to our own detriment to assume that the untried machine can displace the tried and proven horse."[3]

As late as 1939 Rear Admiral Clark Woodward sniffed, "As far as sinking a ship with a bomb is concerned, it just can't be done."[4]

Contrast that unwavering mind-set with the mid-1990s conflict in Bosnia, when some purport that the war was won entirely through the use of airpower.

Similarly, no one really knows the true promise of directed energy today, and it will take years for its true capabilities to emerge and be realized.

And even then, will directed energy ever achieve the same respect that aircraft or nuclear-powered aircraft carriers have today, in being a multipurpose weapon system?

That is not clear. And until then, efforts to transition directed energy to the war fighter will be viewed with suspicion. Warriors want to see a smoking hole in an enemy target and will always ask, "Why use a laser or HPM weapon when a bomb, bullet, or missile will do?"

Another problem with transitioning directed energy occurs when a war fighter wants a new capability—usually he wants it now or maybe in three months at most, but certainly not in two to five years, especially not with today's overwhelming concern in fielding new capabilities to defeat terrorists.

All those involved in the fight against terrorism—from the homeland defender to the soldier in the field—want a new weapon "now." They can't afford to wait the years it takes to transition a breakthrough in science and technology to provide a new leap in capability.

The problem is that if the war fighter keeps coming back to the technologist and asking "what can you give me in three months?" all he'll get is more of the same: something the technologist can take off the shelf and hand to the war fighter.

And if the war fighter only buys a class of weapons that can be fielded in three months or so, he'll never get anything new. At least not a breakthrough in new capability. He may get a slight evolutionary increase in range, power, or number of bullets, but not the revolutionary leap that might be the next "killer app."

This problem keeps surfacing with directed energy, and cannot be overstated: DE may work great in a laboratory environment, under precisely controlled conditions when operated by a cadre of Ph.D.s, but it takes money and time to transition DE weapons into the field, where the weapon can be operated by war fighters who are not mesmerized by the science, but instead who work in what von Clausewitz called "the fog of war."

Part of transitioning a DE weapon includes knowing the context of how the weapon will be used: does the laser or HPM have to work in a 100-degree desert with blowing sand or in a humid jungle and with the ability to punch through dense vegetation? Or does the weapon have to work in both arenas?

Or does the DE weapon have to work in a maritime environment, complete with a hazy atmosphere and sea spray? And must it be mounted on a flatbed trailer, on an aircraft, or on a ship so that it can be wheeled into action as soon as the vessel reaches port?

If there is an overwhelming reason why directed energy weapons will never be used, transitioning this new and untested technology to the war fighter is perhaps the greatest reason of all.

## Directed Energy May Not Be the Most Economical Solution

Remember that scene in *Indiana Jones* when the sword-wielding bedouin threatens Harrison Ford? Instead of going head-to-head with the threat, Ford simply pulls out his gun and shoots the poor man. Ford used the most economical solution—not in sexiness or in cost, but in availability and being able to best accomplish the mission.

Sure, directed energy weapons travel at the speed of light and can do precise damage to a target—but is that always the right thing to do? It may be the sexy thing, but would it be more economical to use a Hellfire missile to accomplish the same mission?

In other words, just as specific weapons are best used in certain situations, DE is not a panacea, a "silver bullet" that can do all things for all warriors.

The issue comes down to identifying what unique mission DE can accomplish. And if that mission exists, is it really worth using DE?

No weapon today can destroy ballistic missiles in the boost phase, and the airborne laser (ABL) appears uniquely qualified for that mission.

Some argue that using a fleet of space-based interceptors to attack ballistic missiles in boost phase would actually work better than the

ABL. But the cost of deploying and maintaining such a fleet would be staggering, and certainly not more economical than deploying a fleet of ABLs.

Although the ABL costs a large amount of money for what the military calls a "high-value asset," there are several missions that require high-value platforms, such as tracking moving ground targets in real time (JSTARS) or directing air battle operations (AWACS).

ABL may be a high-value asset, but like the JSTARS and AWACS, if the mission is critical and cannot be performed by any other means, the mission justifies its cost.

## Directed Energy Tends to Oversell and Underperform

Directed energy has captured the public imagination from the moment Theodore Maiman invented the world's first laser. And soon after Maiman announced his discovery, the military wanted to make a weapon out of it. But actually building a directed energy weapon is tough. As already noted, DE weaponers tend to market their programs optimistically (oversell) and strive for success-driven schedules (underperform).

There are two reasons this happens. The first is that potential capabilities are easy to show on a PowerPoint briefing. These are known as vaporware, when there is no hard data to back up the claims. Anyone can flaunt a concept, especially one based on what is possible and not on what is probable. (More on that later.)

Vu-graphs are cheap, and a sales pitch is easy to make. But until an experiment is performed or until a demonstration is accomplished, a claim is just a claim and not a statement of fact; it merely suggests that something may be possible.

There is a huge difference between something being possible and something being probable. In Chapter 7, I show that although it may be possible to create a region of "antigravity" on earth, in reality it's pretty improbable that it will ever happen.

In many ways it's similar to arguments used by proponents of supersonic transport. It's certainly possible to fly faster than the speed of sound, but building an economically feasibly supersonic transport is not very likely—in other words, it's improbable. Granted, it's improbable because of the business case, but improbable is still improbable.

Similarly it may be possible for directed energy to bounce lasers off orbiting mirrors to strike targets around the world (we'll cover this in Chapter 11), but practically and economically we may never see this happen, again, mostly because of the cost and commitment, not because it's impossible to do.

DE underperforms because building directed energy weapons is tough, as is just about anything having to do with high-energy DE.

Take laser fusion.

The sun and stars create energy through fusing hydrogen atoms to make helium; this fusion process releases enormous amounts of energy that we on earth see as light.

Fusion works. It's a fact, not a pie-in-the-sky idea that will never see the light of day (sorry). It's what makes the sun shine.

Humans have learned to make fusion work on earth in the form of thermonuclear bombs. Creating fusion not only is possible but is a reality—international arms control agreements such as the nuclear nonproliferation treaty are based on it and the Cold War was fueled on its specter.

On the other hand, laser fusion is the noble idea of using lasers instead of explosives (such as in a hydrogen bomb) to compress hydrogen and create fusion. In theory, laser fusion is possible. The massive National Ignition Facility at Lawrence Livermore National Laboratory is constructed to accomplish this very purpose. And I hope my colleagues will succeed, because it will provide critical information for advancing physics and national security.

But several decades ago, accomplishing laser fusion was thought to be just around the corner, in the sense that it would "only" take a kilojoule (a thousand joules, or about 1/40,000 of a stick of dynamite) of energy to fuse hydrogen.

Was it possible? Yes. In theory lasers should have been able to take advantage of a phenomenon known as the rocket effect when imploding a tiny fusion target. At least the calculations indicated it was possible.

But was it probable? The answer was a resounding no.

The amount of energy turned out to be wrong by about a factor of 1,000, and a host of other problems popped up as well.

Through a success-driven schedule that did not adequately account for high-risk problems or build in the time and money to fix those unknown problems, the initial laser fusion proponents oversold their idea and failed because of underperforming. In this case, the laser energy needed to achieve fusion energy underperformed by a factor of 1,000, a lesson other areas of directed energy would do well to heed.

## How Directed Energy
## May Be Used Against Us

The biggest problem with DE is that some of the unique features of directed energy—speed-of-light engagement, precision engagement (lasers), broad area coverage (HPM), circumvention of gravitational ballistics, decreased logistics tail, and the deep magazine—might also be used against us, allowing an asymmetric attack to be mounted against our own capability.

For example, the United States and its allies employ the most technologically sophisticated weapons in the world, exploiting computer chips, intricate electronic circuits, and software that allow them to overwhelm their enemies. However, this unprecedented reliance on electronics has resulted in a virtual Achilles' heel that adversaries might exploit by employing HPM attacks against our own weapons.[5]

From a vulnerability perspective, the good news is that since the United States may be a decade or more away from fielding its own HPM weapons, our enemies would take even longer to gain this capability. As we'll see later in this book, fielding a robust HPM weapon is tough if not impossible. And if the United States is having difficulty solving the scientific problems that plague HPM, then others are just as strapped, if not more so. Thus our electronic weaponry may not be vulnerable for some time yet.

On the other hand, again from a vulnerability point of view, because knowledge increases so rapidly, unforeseen breakthroughs in science and technology may result in a revolutionary increase in HPM capability for not only the United States but our adversaries as well.

But the worst news is that HPM effects are not limited to negating sophisticated military weapons. The entire U.S. infrastructure—banking, telecommunications, transportation, the power grid (both electric and gas), and the food and water supply—may be vulnerable as well.[6]

It is no secret that as the most technologically sophisticated country on the planet, the United States is heavily dependent on modern electronics, and increasingly so in sophisticated computer chips and software. And as the father of high-power microwaves warned in advocating HPM, what makes our enemies susceptible, makes the United States even more vulnerable. As Dr. Bill Baker cautioned, "The smarter the weapon, the dumber HPM can make it."[7] This vulnerability extends beyond the circuitry found in sophisticated weapons to include the U.S. infrastructure due to its reliance on electronics.

Nonelectronic elements of the U.S. infrastructure are vulnerable to other types of directed energy as well. Witness the near panic that resulted from airline pilots being "lased" by not much more than pocket lasers in early 2005. The FBI is aggressively pursuing the laser-dazzling problem, and it should not be surprising that military countermeasures such as laser-safe eyeglasses may someday be used by commercial airlines.

The relative softness of the U.S. electronic infrastructure in banking, telecommunications, and utility control may prove to be our nation's Achilles' heel. Imagine terrorists taking down Wall Street with an HPM weapon—a idea considered outlandish by fiction publishers in the early 1990s[8]—or the banking industry being decimated if its computers inexplicitly crashed.

The upshot is that the U.S. arsenal of sophisticated military weapons may not be the only target of directed energy weapons that adversaries may try defeat. The very core of what makes up American culture—our water supply, electric grid, monetary transactions, air travel, and just about any other feature dependent on electronics—may be at risk in the future.

## The Solution

Directed energy holds great promise, from neutralizing ballistic missiles to providing a means of applying nonlethal force that gives the war fighter an option in between shouting at someone or shooting him.

There are historical precedents for inserting advanced science and technology into the battlefield, to radically extend both the range and the lethality of the weapons of war. Although military effectiveness will be increased by the use of directed energy, paradoxically the lethality will be *controlled* by the fact that DE is not a weapon of mass destruction. Rather, it is almost infinitely precise, allowing quite literally a surgical application of force.

And even better, this accuracy can be achieved at the speed of light. So there is no wonder why war fighters and policymakers are excited about the vast potential DE promises.

But there are major problems ahead for directed energy, problems that cannot be dismissed or easily overcome. And in the future the United States will have to face its own vulnerability to DE weapons if they are ever employed by our adversaries.

Transitioning directed energy to the war fighter and protecting our infrastructure requires a massive investment in time and funding. It involves technically maturing the DE science and technology base, including directed energy infrastructure—sources, beam control, target acquisition, pointing and tracking, all of which are discussed in Chapter 5. It also involves educating the war fighters by allowing them to experiment with directed energy weapons so they can fully integrate DE into the battlefield.

Directed energy's capability must be realistically assessed and sub-stantiated by demonstrations. And while this happening, it must not be oversold, and it must perform as promised.

And this takes time, commitment, and money, for it cannot be done overnight.

The next chapter contrasts lasers and high-power microwaves, showing that there are similarities—such as they both travel in straight lines—as well as differences—in range, power, and application. But despite these differences, they're both made up of photons and they both travel at the speed of light.

# 5

# The Difference Between
# Lasers and High-Power Microwaves

WE'VE SEEN THAT LASERS AND MICROWAVES have more in common than what we'd first suspect. In fact, they're really just slight variations of the same thing—they're different wavelengths of electromagnetic radiation, or, equivalently in a quantum worldview, they have different photon energies.

Electromagnetic waves have characteristics that define how fast they spread out (diffraction), how far they propagate (range), and how they interact with various types of matter. These characteristics all arise from variations in wavelength (or photon energy). To understand why lasers and microwaves diffract, propagate, and interact differently, we first have to understand these different characteristics.

## Diffraction

When electromagnetic waves leave the system that generated them (an amplifier, a laser, a microwave cavity, the final optics, etc.), they diffract or bend around the system's physical exit point.

This phenomenon was first proposed by the English physicist Robert Hooke in 1665 and later formalized by the Dutch scientist Christian Huygens in 1885.

The diffraction of waves is explained by Huygen's principle, which results from regarding every point on a wavefront as a new source of waves. We can visualize this by considering a series of parallel waves impinging on a wall that has a small opening. Obviously the waves hitting the wall are going to be reflected back in the opposite direction. But what about the waves that miss the wall and go through the small opening?

Huygen's principle dictates that each point on the wavefront may be considered a new source of waves. Thus new waves will squeeze through the small opening and squirt past the barrier.

But instead of new plane waves being created past the barrier, if the opening is very small with respect to the wavefront's wavelength, that source of new waves will create circular waves, not parallel waves.

The curvature of these new waves is called diffraction, and it depends on the size of the opening and the wavelength of the initial waves.

As in Chapter 2, if equations make you nervous, simply skim over the text; you don't need to work through the math to understand the phenomenon. I am including the equations to make this book complete.

Figure 5.1 shows an example of plane (parallel) waves impinging on a solid wall with a small hole. Note that the waves coming in from the left are all parallel to one another. As a rule, the bigger the wavelength ($\lambda$) is, compared to the opening (d), the more the wave diffracts. In general, the diffraction is proportional to $\lambda/d$.

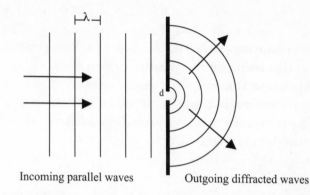

Incoming parallel waves          Outgoing diffracted waves

Figure 5.1    Diffraction of parallel, plane waves of wavelength [λ] through an opening d. The plane waves are assumed to come from a source an infinite length away (so that the spherical nature of the waves at infinity are planar). The opening d is assumed to be much smaller than the wavelength (d<<λ). Note the spherical nature of the diffracted waves above, after exiting the opening. This is due to Huygen's principle, which postulates that every point on a wavefront is a point source of a new wave.

## How Diffraction Affects a DE Beam

What does diffraction have to do with directed energy? Remember that the only difference between lasers and microwaves is their wavelength.

Since diffraction is proportional to $\lambda/d$, and since laser wavelengths are 10,000 times smaller than microwaves, for the same size aperture, lasers diffract 10,000 times less than microwaves, or they "spread out" 10,000 times less.

Since lasers spread out 10,000 times less than microwaves, they can deposit the same amount of energy a distance 10,000 times farther than microwaves.

Because microwaves diffract so much more than lasers, someone may ask, why even use microwaves when you can use a laser?

One reason is that you may want to exploit this large difference in diffraction. Suppose you want to irradiate a large number of targets at the same time. Since at the same distance a laser beam spreads out 10,000 times less than a microwave, it might make sense to irradiate the area with microwaves—blanket the area all at once. High-power microwaves give us the ability to do this much better than lasers.

Just because microwaves diffract (spread out) more than lasers doesn't mean that lasers are "better" than microwaves. It all depends on what you are trying to accomplish, and beam diffraction is only one parameter that you must consider.

## Range

Range is a measurement of the distance at which a weapon is effective. As in diffraction, just because one weapon is more effective over a longer distance than another doesn't make that weapon any better. It really depends on how you use the weapon, or in military terms, how the weapon is employed.

For example, is a shotgun a better weapon than a highly accurate, long-range rifle? The question is meaningless because the answer is: It depends. It depends on what you're trying to accomplish and under what conditions you're shooting.

A shotgun is great for targets that are spread out or are relatively close. Flushing quail on the moors while you're tramping through dense brush begs the use of a short-range weapon that can cover a wide area.

On the other hand, a sniper zeroing in on a target 400 yards away needs a highly accurate long-range rifle to bag his prey. A shotgun would startle the target and ruin the shot.

So if range is not an indicator of effectiveness (remember the definition of military effectiveness supplied in Chapter 3), then why is range important for directed energy? Especially since directed energy, traveling at the speed of light, goes around the world seven times in a second.

But isn't the range of directed energy infinite? Yes, at least in relative terms. After all, in the time it takes a .22-caliber bullet to travel a kilometer (roughly half a second), directed energy travels 150,000 kilometers. That's a heck of a lot longer range than any other weapon!

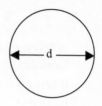

Figure 5.2    A spot size of diameter (d) from a beam of electromagnetic radiation hitting a planar (i.e., two-dimensional sheet) target.

So why is range even important in directed energy? And if it is important, then what has the longer range: lasers or microwaves? And since they're both different aspects of light, does it even matter?

Good questions, all. So let's put range in terms of directed energy.

First, remember the diffraction phenomenon described above. Diffraction—how much light spreads—depends on the wavelength and the size of the aperture. Diffraction increases when we increase the wavelength, and that decreases the amount of energy per unit area that we are able to put on a target. Thus, if you want to deposit a lot of energy at a distance, you want very little beam diffraction. So at first blush it seems that less beam diffraction means a longer range.

But before we draw any conclusions, let's look at this again.

When a beam of electromagnetic radiation (visible light, microwaves, ultraviolet radiation, radio waves, infrared rays, etc.) hits a target, the beam illuminates an area called a "spot size." If the beam is perpendicular to a planar target, the spot size is circular.

For example, when you switch on a flashlight, you project an oval patch of visible light on the wall. The oval patch is called the spot size. As you move the beam perpendicular to the wall, that oval shape becomes a circle, as shown above. But regardless of the shape—oval, circular, or whatever—the radiation pattern is called a spot size. And lasers, microwaves, and every type of electromagnetic device create a spot size of radiation.

So what does this tell us and why is it important?

Since diffraction depends on the wavelength of light (remember that a smaller wavelength means less diffraction), then the spot size of light

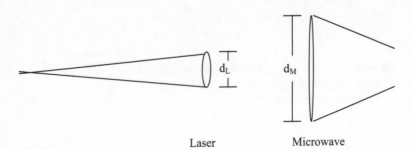

Figure 5.3   The spot size from a laser ($d_L$) is typically smaller than the spot size from a microwave ($d_M$).

($d_L$, from a small wavelength) should grow less than the spot size of a microwave ($d_M$, from a larger wavelength).

Thus a laser can project its energy into a tight spot farther than a microwave weapon; or for an equivalent distance, $d_L$ is smaller than $d_M$.

But is this good or bad?

Again, just as with shotguns or long-range rifles, the answer is: It depends.

Let's look at some examples.

The airborne laser (ABL) will want to stay as far away as it can from enemy territory, yet deliver as much destructive laser radiation as it can against an enemy missile. Thus the ABL wants to use a tightly focused beam whose spot size doesn't spread out very much as the laser energy is projected across hundreds of kilometers of airspace.

On the other hand, suppose that the military has finally overcome all significant technical hurdles (power, atmospheric breakdown, large antenna, etc.) for a high-power microwave (HPM) weapon and has put this device on the battlefield.

If commanders could instantly debilitate every electronic weapon—missile, airplane, communication, firing computer, command control, and intelligences node—on the battlefield, then they would not hesitate to use HPM to help win.

Table 5.1    Contrast Between Lasers and HPM

|  | *Lasers* | *Microwaves* | *Difference* |
|---|---|---|---|
| Speed: | Speed of light | Speed of light | None |
| Trajectory: | Line of sight | Line of sight | None |
| Range: | Hundreds of kilometers | Hundreds of meters | HPM range is 10 to 100x less than lasers |
| Power: | Megawatts | Gigawatts | HPM power is 1,000x greater than lasers |
| Wavelength: | Short (100s of nanometers) | Long (mm to cm) | HPM waves are 10,000x longer than lasers |
| Beam: | Narrow | Broad | HPM spreads out 10,000x more than lasers |
| Military use: | Precision | Large area | |
| Target: | Equipment and people | People, and in the future, electronics | |
| Lethality: | Burn surface, dazzle | High voltage, EM field; heat | |

Conversely, commanders would not try to disable or destroy these critical components one by one with a laser. Why not? Because focusing on each target would take too long. And lasers can't punch through smoke and clouds, "the fog of war," while microwaves are impervious to that type of scattering.

So what does this tell us, and how does a weapon's range mix into the way directed energy weapons are used?

Once the utility of directed energy has been proven, through ABL, MTHEL, ATL, or even ADS, DE will be treated just like every other weapon in human history. If it is the best system for a given situation, it will be used.

Air-to-air missiles are not used by soldiers storming a redoubt, hand grenades are not used to shoot down a enemy fighter jet, and bayonets are not used against tanks.

Weapons are used against a specific set of targets for which they have been optimized to defeat.

Directed energy will be no different.

If a target is hundreds of kilometers away and needs to be destroyed in near real time, then lasers may be the best solution when employed with relay mirrors.[1]

In the future, if electronics on a battlefield need to be neutralized, then HPM may be the solution.

And just as in controlling a crowd from a long distance or assessing the intent of enemy combatants, then millimeter waves in the form of the Active Denial system (ADS) may be used. Table 5.1 contrasts lasers and microwaves.

But despite their differences, as electromagnetic waves, directed energy weapons

- Are unaffected by gravity (i.e., they travel in a straight line)
- Are capable of a full spectrum of graduated effects, "from toast to roast": deny, disrupt, degrade, and destroy
- Cause minimal collateral damage

And, most importantly, they do all this at the speed of light.

# 6

# Prophets of the Revolution: High-Energy Laser Weapons

YOU MAY BE SURPRISED TO DISCOVER that the laser has a contentious past. To this day there are legal questions as to who was directly responsible for this fundamental discovery. As with all world-changing breakthroughs, however, no one person was solely responsible. The laser actually resulted from 100 years of fundamental scientific advances.

In 1856 James Clerk Maxwell laid the foundation for the laser when he combined the then-disparate fields of electricity and magnetism through his now famous Maxwell's equations. Maxwell's equations describe the nature of light, showing that the different manifestations of the electromagnetic spectrum—radio waves, microwaves, heat, visible light rays, and gamma rays—are similar and vary only in wavelength or, equivalently, in frequency or energy.

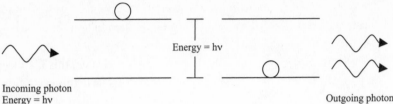

**Figure 6.1**    On the left-hand side, an incoming photon (bundle of energy) of energy E interacts with an atom that is in an excited state of energy E. On the right-hand side, the atom has given up its energy as a photon of energy E, traveling in phase and in the same direction as the incoming photon. When photons have equivalent direction and phase they are called coherent.


Sixty years later, Albert Einstein's invention of what is now known as the Einstein coefficients allowed scientists to construct a theory of the stimulated emission of light. This was another fundamental advance in the path toward the laser.

An easy way to understand stimulated emission is to imagine a light wave interacting with an atom or a molecule. Suppose the light wave consists of an infinitesimal amount of energy, so that we have only one photon of light. Crudely put, if the atom or molecule has the same amount of energy as the incoming light wave—the atoms or molecule is in an "excited state"—a photon of equivalent energy will be emitted resulting in two photons. Figure 6.1 shows this process, known as stimulated emission. If the new photon is emitted in the same direction and has the same phase as the incoming photon, this is known as coherent emission.

Now imagine those two coherent light waves interacting with two additional atoms or molecules that are in an excited state. A photon from each atom or molecule will be emitted, resulting in four photons, all having the same energy traveling in the same direction and phase.

You can imagine how quickly the number of photons grows, especially since a typical laser median has billions upon billions of atoms or molecules in an excited state, ready to emit photons.

When the atoms or molecules are in an excited state, this is called a population inversion. Atoms or molecules don't have this excitation energy; normally they are in a "ground" or unexcited state. They become energized when an outside power source is introduced to the system through the creation of a chemical reaction (such as what happens in a COIL laser), an electric discharge (such as in an electric discharge laser), a flashlamp (such as the ruby laser), or some other excitation mechanism. This energy boosts the atoms or molecules from a ground state to an excited state, creating a population inversion.

The last significant step for inventing the laser occurred when coherent radiation from stimulated emission was documented in 1953 by H. Motz and a team of collaborators. Motz used a radio frequency (RF) linear accelerator to drive an electron beam through a series of alternating magnets, resulting in coherent radiation.[1]

Now that stimulated emission and coherence had been observed, all that remained was putting it all together. This set the stage for the invention of the laser.

The microwave equivalent of a laser, the so-called maser, was invented in 1958 by Arthur Schawlow and Charles Townes. The two Bell Labs researchers patented the device. Although they did not actually construct a laser, in a December 1958 paper published in *Physical Review* they extended their theory of the maser to the optical regions of the electromagnetic spectrum.[2] Many point to this event as the true birth of the laser; all that remained was to actually build one—merely an engineering problem to some.[3]

As it turned out, the solution to the engineering problem Schawlow and Townes encountered in building a working laser was not as evident as the solution to building a maser.

In a maser, microwaves bounce back and forth in a conducting cavity. The microwaves grow in intensity as they are reflected at either end of the cavity by a metal mirror. Like a train engineer continuously stoking a growing fire, with each pass the microwaves grow in intensity. If there were no way to siphon off the microwave energy, the microwaves would escalate in intensity. However, the microwave energy was bled off in a small side compartment known as a waveguide and projected out of the cavity.

In Schawlow and Townes's paper, it was not obvious how light could be siphoned off from inside a laser cavity. A wave structure for light similar to that of a microwave's wave structure would be at least 10,000 times smaller, and building micron-size cavities was not practical.

The problem was solved on May 15, 1960, at the Hughes Research Laboratory in Malibu, California, when Theodore Maiman generated the world's first laser beam using a small crystal of ruby.

Instead of siphoning light off through a waveguide, Maiman used a partially reflecting mirror at one end of the cavity, allowing a small percentage of the laser light to seep through the mirror.

With a partially reflecting mirror, the engineering problem was solved and the news of a working laser spread through the media like wildfire.

The invention quickly caught the military's attention, as it was looking for the next big technological breakthrough to give it an advantage in the Cold War.

Years before, first the atomic bomb and then the hydrogen bomb had provided the technological edge to keep the Soviet Union at bay; miniaturizing the weapons to be lofted on highly accurate ballistic missiles gave the United States and its allies another edge in the arms race.

This strategy of investing in science and technology to leapfrog national security threats (known as a technology-intensive strategy) was needed in the Cold War to counter the more massive manpower-intensive strategy of the Soviet bloc.[4]

Futurists projected that the military could use lasers as infrared countermeasures for air-to-air missiles, in underwater surveillance, as laser radar, for active imaging (covertly illuminating a target in the dark), for illuminating the battlefield with a specific laser frequency that only those outfitted with special goggles could detect, in tactical fighter jets, and as high-power strategic laser weapons.

Even though the laser had only demonstrated a minuscule amount of power and would need to increase that power by over a thousand million times to be useful as a weapon, the military was giddy about the possibilities. In that context Maiman was cautiously optimistic about the significance of his groundbreaking research when he stated, "The laser is a solution looking for a problem."[5]

In the 1960s, low-power lasers were of little use to the military. Laser designators for precision-guided weapons and laser gyros were a novelty; later, GPS would replace most of the applications for laser targeting and navigation. The future for lasers seemed bright, but the military needed a lot more power to produce a strategically significant laser weapon, and that would take decades to achieve. The military was less interested in low-power lasers and focused its research instead on reaching high levels of power.

In contrast, low-power lasers were used by industry to solve previously intractable problems. Medical lasers were used to perform delicate eye surgery; miniature, low-power solid-state lasers were invented to read bar codes, speeding supermarket checkout lines and revolutionizing inventory control; and intermediate-power lasers were designed for a myriad of uses from cutting through steel to drilling precision holes in baby bottle nipples.[6] Commercial low-power lasers are also used in CD and DVD players, optical fiber communication, remote sensing, weather sensing, TV and the entertainment industry, reprographics, tissue cutting, photodynamic therapy, circuit repair and marking, lithography, and heat treatments.[7] The market for low-power lasers soared, and today investment by the telecommunications industry alone exceeds $4 billion.[8]

The military set out to rapidly advance the state of the art in high-energy lasers (HEL), and each of the services established its own programs to increase the laser's power and prove the military utility of fielding an HEL weapon. The services focused on three technology areas that needed significant advances in order to achieve an HEL: beam quality, beam control, and beam power.

A working laser system involves complicated trade-offs between size, weight, and operational complexity (including reactant storage and condition, exhaust handling, cavity pumping, weight of the hardware systems, logistics, requirements for deployment, operations, atmospheric conditions, laser power, etc.).[9] For simplicity and to keep this discussion manageable, we'll consider only beam quality, control, and power.

## Beam Quality

Increasing the power of a laser is only part of the problem of fielding a laser weapon.

Once a laser beam is generated, it has to be of the best possible quality so it will have the highest intensity when it reaches a target. Some lasers, such as COIL (chemical oxygen-iodine laser) and FEL (free electron laser), have inherently high beam quality, but they have not yet proven the ability to produce the megawatt class of power needed for a strategic HEL weapon (although the COIL laser should reach that milestone in 2006, when the airborne laser tests its COIL modules). Other lasers such as DF (deuterium fluoride) have achieved megawatt-class powers, but they have relatively poor beam quality.

Compared to the other problems in producing a HEL weapon, generating a high-quality laser beam was considered by some in the military as a science project, something that could be performed at a top-notch research university. For example, advances in solid-state laser beam quality continue today, driven by the telecommunications industry for applications using fiber optics.

Once a beam has been generated with the highest-possible quality inside the laser cavity, numerous nasty things can happen when the beam leaves the cavity and propagates to a target. Some of these things can be controlled, but some cannot.

Most laser experiments are conducted in carefully controlled conditions in a laboratory. There, the researcher can propagate the beam inside a vacuum or in very low pressures of pure gas. This provides a known, measured environment where the properties of the beam can be observed and corrected.

In the real world, the atmosphere is a jumble of different gases, including nitrogen and oxygen, existing in an ever-changing environment. Worse, even in very dry areas, the air is roiling with water, dust, pollen, and all sorts of particulates. The atmosphere fluctuates with differing densities of these constituents, and varies in temperature from over $100^0$F to $-50^0$F. Layers of differing conditions exist, and random conditions occur through turbulent mixing.

In this dynamic, real-world environment, the laser beam expands, diverges, scatters, reflects, and even changes the atmospheric conditions along the path it takes.

Some frequencies of light—in the terahertz region (from 300 gigahertz to 100 terahertz), and X ray and gamma radiation—are absorbed by the atmosphere and don't propagate very far. When other frequencies finally reach a target through this real-world atmosphere, rarely are they of the same high quality needed to produce a weapons effect.

But even with these formidable problems, there are various ways to circumvent these curve balls nature is throwing at high-energy lasers.

One way is to *tune* the laser so that it avoids some of these undesirable characteristics of the atmosphere. For example, the HF laser (2.8 microns) is highly absorbed in the atmosphere. But when the DF's hydrogen atom is replaced with a deuteron (a hydrogen atom with an additional proton, making it a DF laser), DF laser light lasing at 3.8 microns punches cleanly through the atmosphere.

Another major problem—beam breakup by a turbulent atmosphere when the laser is projected hundreds of kilometers through the atmosphere—can be solved by using a technology called adaptive optics (or

AO as it is known by laser researchers). This is a beam control problem and is covered below.

## Beam Control

An HEL beam must be steered toward a target, and that means developing accurate optics, typically in the form of painstakingly aligned, high-quality mirrors. If the mirrors are not exactly lined up to within very small tolerances, the high-power beam could miss the mirror, graze past, and blast anything in its path—equipment, walls, or even people.

What's more, in a high-energy laser the optics must endure very high power levels, approaching a billion watts over every square centimeter of the mirror.

In low-power commercial applications, such as steering a laser beam in a DVD player, the lasers are billions of time less powerful than an HEL; in a high-energy laser, the same delicate mirrors would be instantly vaporized if the same low-power technology were used to steer the beam.

Controlling the laser beam is more than just building high-quality mirrors and ensuring they are accurately aligned. First, the beam control system must extract the laser from the cavity, which may be an oscillator or an amplifier.[10] During this extraction phase, the beam control system (BCS) must ensure that the beam does not lose energy (i.e. "lossless"), or decrease in quality. Further, it must accurately line the beam up with the optics and mirrors in the BCS. This series of optics and mirrors—the optics train—is painstakingly difficult to align. Low-power lasers are typically used to ensure that the optics train steers the beam to the correct location.

As the laser propagates through the optics train, it may be changed or morphed into an optimum shape that can be directed out the final aperture, which is usually a telescope that focuses the beam onto a target. In addition, the beam may be predistorted to take into account any aberrations in the atmosphere.

As already noted, these aberrations are caused by turbulence in the air. The atmosphere changes in a small amount of time known as the hydrodynamic timescale, typically several hundred times every second. In the BCS, the laser beam can be cleverly predistorted so that as the laser travels toward the target, the beam actually distorts back to nearly the same pristine condition it had when it left the laser cavity. Since the atmosphere changes hundreds of times a second, this predistortion must be made hundreds of times a second as well.

This technique of predistorting the beam is called adaptive optics (AO), first proposed in 1953 by Horace Babcock for compensating astronomical images in real time to create higher resolutions in telescopes.[11] AO was refined and eventually applied practically for the first time in a highly classified Strategic Defense Initiative program dating back to the early 1990s.[12]

For an HEL weapon such as the ABL, a low-power laser is used to illuminate a target, and the beam is distorted by the atmosphere when it reaches target. That low-power laser is reflected back from the target to the ABL, where the distorted beam is analyzed. Now that the ABL knows how the reflected beam has been distorted by the atmosphere, it can quickly calculate in millionths of a second what it would take to predistort the high-energy beam.

The outgoing HEL beam is then distorted by the conjugate (roughly, the opposite) of the reflected low-power beam's phase by bending a flexible mirror over 2,000 times a second. This distortion occurs many times faster than the atmosphere can move, so in effect, as the predistorted HEL beam propagates to the target, the atmosphere doesn't change. For that very short period of time (submilliseconds), the atmosphere appears frozen, and the HEL distorts to a near perfect beam at the target.

This technique does not work well in extremely turbulent air, and it is not always useful for short-range (up to 20 km) engagements. A more detailed discussion of criticisms of adaptive optics, including concerns by the American Federation of Scientists and the Defense Department's own watchdog organization (PA&E, Program Analysis and Evaluation), is presented in Chapter 10.

While the BCS is shaping and preparing the beam to be directed toward the target, it is simultaneously conducting another, equally important phase. It must acquire and accurately track the target. The farther the target from the laser, the more precise the BCS must be to ensure the beam is successfully held on the aim point.

For a ground-based, stationary laser, this is a straightforward problem. Graduate students in astronomy have tracked low-flying satellites for years, and there are many commercial businesses that produce highly accurate, stable trackers for the astronomy community.

For lasers on a moving platform—ground, sea, air, or space—this problem quickly escalates in complexity. For the ABL, not only is the laser moving relative to the target, but in an airplane the laser is being buffeted by atmospheric turbulence as well as the airplane's own low- and high-frequency vibrations that come from the engines and flight-control systems.

As complex as the BCS in an airborne laser may be, one can only imagine the complexity of constructing one for a space-based laser (SBL)—especially since other, relatively "quiet" concepts such as solid-state lasers are years behind the power levels needed for weapon as compared to chemical lasers. Chemically based SBLs will be run like a rocket, creating an even more challenging environment for the beam control system to operate.

With these complex requirements of shaping the beam, ensuring the precise and lossless transit of laser radiation through the optical train, acquiring and tracking the target, then accurately holding the beam on its aim point, to many, the beam control is the "long pole in the tent"— the hardest problem in producing a strategic high-energy laser weapon system, especially when that weapon is in the real world, not a pristine laboratory environment.

And the danger for HEL systems is that this problem may indeed prove insurmountable.

## Beam Power

Conventional wisdom says that increasing laser power is simply a "scaling" problem. That is, compared to building a robust beam control system on a space-based laser, achieving high laser power was viewed as less difficult.

If a problem can be scaled, that typically means the power can be increased by adding more energy. Think of water flowing out of a garden hose. The flow increases when you turn the handle and force more water into the hose. In a crude sense, the amount of water flowing through a hose is scalable.

Similarly, to drive a nail deeper in a plank of wood, all you need to do is to use a bigger hammer—a sledgehammer works better than a ball peen. But if you try to drive that same nail through a slab of tungsten, the nail will bend or its head will smash. Past a certain point, the problem is not scalable. Instead, you need to use an entirely different approach to drive a nail through tungsten.

In much the same way, in the 1960s the various low-power lasers being invented were all scalable to some degree. For solid-state lasers, adding more crystal slabs increased the amount of gain media, and the power increased correspondingly. By 1964, in the newly invented $CO_2$ gas laser, increasing the electric discharge arcing through the laser cavity usually resulted in more power.[13]

However, without exception, no matter how much more power researchers tried to wring out of lasers, eventually a fundamental limit would be reached and no additional power could be wrung out of the system. Sometimes the laser would even stop working.

Scientists realized that if high-energy lasers were ever to become possible, an entirely different method of extracting power would have to be invented.

## Supersonic Expansion

Shortly after the laser was invented, two Soviet scientists proposed that rapidly cooling a laser medium might produce a large population inversion.[14] Using this unconventional approach with gas lasers, some speculated that this technique might be used to circumvent the scaling limit and produce a high-energy laser.

Researchers knew that the laws of thermodynamics dictated that one way to rapidly cool a medium was to expand a flow of hot gas through a supersonic nozzle, much the same as what happens in a rocket engine. In 1966 this technique was successfully demonstrated in a 135,000-watt (135-kilowatt) gas dynamic laser operated at the Avco Everett Research Laboratory.[15]

Supersonic expansion is now used in every major gas laser system, and power levels have increased with this technique to produce megawatt-class lasers.

## The Pursuit of High-Energy Laser Weapons

In the early 1960s, the possibility of fielding a Buck Rogers "blow the enemy to hell" beam weapon lured the military to pursue speed-of-light technology. Although the difficult issues of beam control and beam quality had not yet been realized, optimism reigned. As shown in the previous section, increasing the laser power seemed to be the biggest problem and even that appeared straightforward.

During the heady 1960s, anything seemed possible.

Scientific success with atomic weapons fueled this unbridled optimism. Twenty years earlier, although needing many tens of kilograms of plutonium for an atomic bomb, the Manhattan Project had been started after only micrograms had been produced in university laboratories. In essence, under immense wartime pressure to build the world's first atomic bomb, the country was betting its scientists could somehow produce plutonium in quantities tens of billions of times greater than had ever been produced in the lab.

And they succeeded.

Physicists were heralded for their achievement. This set the standard and the expectation that science and technology could save the day.

This expectation was given further impetus by the fledgling field of rocket science in the 1950s. With a failure rate approaching 60 percent, in a span of a decade, missile technology advanced from lifting 10-pound payloads a few miles off the ground to throwing tons of material intercontinential distances.[16] Again scientists were held in high esteem as ICBMs became commonplace.

Faith in science and technology grew as supersonic fighters roared through the skies when, just years before, breaking the sound barrier had seemed impossible. With the invention of the transistor, vacuum tube technology became obsolete, and electronic devices from radios to radars proliferated at a head-spinning pace.

With this technological shock front rolling across our culture and changing the world, the military had no doubt that scientists could increase laser power enough to eventually field a laser weapon.

The Defense Department began to take an interest in speed-of-light weapons shortly after the invention of the laser. ARPA[17] (Advanced Research Projects Agency) and the air force funded the earliest laser programs, primarily motivated by strategic defense.[18] Since the first successful ICBM launch, there has been no defense against a nuclear weapon lofted from thousands of miles away, and the problem has been compared to trying to stop a rifle shot by hitting it with a bullet.

Early antimissile programs such as Nike-Hercules relied on detonating an atomic bomb close to an incoming ICBM-lofted warhead, creating a nuclear firestorm 10 or more miles above the surface of the earth that would kill incoming warheads, and hopefully create less destruction than an enemy warhead detonating on the ground. This option made many people uncomfortable, and eventually the Nike-Hercules program was scuttled. There was no technological solution for stopping an ICBM, and the nation relied on the uneasy policy of détente and mutual assured destruction (MAD) to hold the world's nuclear trigger finger at bay.

With a lack of science and technology solutions, the military envisioned high-energy lasers as the silver bullet for antiballistic missile defense. Powerful ground-based lasers might be able to stop incoming nuclear warheads, providing an impenetrable shield around the United States. Others saw these lasers being used to destroy enemy satellites, prevent foreign eyes from surveilling our nation, or shoot down bomber aircraft as they fly toward America.

Later the army and the navy became interested in lasers for tactical applications such as air and ship defense to instantly stop threats. Traditional ways of defending army encampments and ships—antiaircraft gun batteries—are too slow to stop low-flying supersonic cruise missiles that are becoming part of an enemy's arsenal.

In the 1970s, R&D funding for these applications approached $100 million a year, with the majority of the funding going to the air force Airborne Laser Laboratory, a demonstration to prove the feasibility of shooting down air-to-air missiles using a $CO_2$ laser on a NKC-135, the military version of a Boeing 707.

But the high-energy laser story involves more than military research. Industry has played a major role, largely as a result of a downswing in the space industry. Back in the late 1960s, now that humans had realized President Kennedy's dream of placing a man on the moon in a decade, spaceflight became routine and the nation turned its eyes to other priorities.

And that led to exciting opportunities matched only by those chronicled in science fiction.

## Industry Discovers High-Energy Lasers

Dr. Joe Miller was an early high-energy laser pioneer, and he was typical of the industrial workers who moved into this esoteric field.[19] Like the majority of government researchers, Joe did not have a background in lasers, but his Ph.D. in engineering allowed him to come up to speed quickly.

As was typical of nearly all early laser engineers, Miller had been working on the Apollo missions during the heady 1960s, when the roaring space program swept up nearly all the science and engineering talent in the country. Not having a background in lasers didn't hurt the companies that were pushing into this new high-tech territory. In fact, it might have been to their advantage, since they were not preprogrammed to look for typical solutions for problems with HELs.

When you're stuck on a problem, the best person to help you out is usually someone off the street who is not immersed in your own field, and can stand back and look at your problem in a new light. Some of the most revolutionary advances are made in this manner, so Dr. Miller's lack of formal laser training not only didn't hurt him but was probably one of the reasons for his future success.

Similarly, Major Don Lamberson, who finished his Ph.D. studies at the Air Force Institute of Technology in what amounted to nuclear effects, was tapped to lead the air force laser program. Later, Lamberson headed the airborne laser laboratory and was promoted to a two-star general.

In 1969, Dr. Miller worked at the TRW Space Technology Laboratory in Southern California, now a part of the sprawling Northrop Grumman industrial empire. During that time the Aerospace Corporation, under contract to the air force Space and Missile Center in Los Angeles, was working with a new chemical laser known as a hydrogen fluoride (HF) laser, powered by an electric arc that propagated through the HF. This 1-kilowatt laser caught the attention of the military, as it was 100,000 times more powerful than Maiman's first ruby laser, which had shocked the world only 10 years before.

As the Apollo program wound down, TRW, along with other companies such as Rockwell, actively sought new, revolutionary high-technology areas in which to invest. These companies had spent large capital acquiring highly trained talent to pull them through the exhilarating years of the Apollo program. In the late 1960s, technology seemed to be the solution for a majority of the world's problems.

Identifying an opportunity to use its highly trained scientists, engineers, and technicians in an exciting new field, TRW invested a portion of its own research money (known as IR&D, or internal research and development funds) to build a 1-kilowatt combustion-driven HF laser at its facility in San Juan Capistrano, California.

The concept championed by Miller and his group was that a high-energy laser needed a tremendous increase in its initial energy in order to achieve powers greater than a kilowatt. It was generally recognized that a strategic, weapons-class laser would need on the order of a megawatt to be a useful military weapon. A megawatt was a thousand times more powerful than a kilowatt, but that did not deter the researchers; after all, Aerospace had achieved increases of 100,000 times over the first ruby laser when it demonstrated its HF laser.

TRW realized that even if laser efficiencies were 100 percent (an impossible number to obtain), electric arc discharges would never supply enough energy to power a weapons-class HEL. Problems such as gas breakdown and beam instabilities would prevent a megawatt-class arc-discharge laser from ever working.

On the other hand, research in supersonic flow performed during the Apollo era had proven that immense quantities of energy could be transferred to a lasing medium during the combustion (and subsequently a supersonic expansion) phase.

The TRW 1-kilowatt combustion effort succeeded, and two short years later in 1971, it proposed to DARPA (Defense Advanced Research Projects Agency) a plan to build a 100-kilowatt laser based on the same technology. TRW's so-called BDL (baseline demonstration laser) caught the eye of the government, which was considering an air force proposal, led by Colonel (Dr.) Pete Avizonis and Lieutenant Colonel (Dr.) Don Lamberson. Avizonis and Lamberson proposed building a 100-kilowatt laser at Edwards Air Force Base, but a decision was finally made to build it at TRW's facility.

After that decision was made, Major John Rich (who later as a full colonel both commanded the Air Force Weapons Laboratory and headed its laser directorate) queried why TRW was able to come in with such a reduced cost and schedule (do it cheaper and faster), as opposed to the air force proposal.[20] The reason was that TRW had proposed constructing the 100-kilowatt laser at its San Juan Capistrano facility, where it could work 24 hours a day, use nonunion labor, avoid driving the 40-mile one-way trip to the Edwards site (as did the air force personnel), and, most importantly, use an optimized nozzle configuration for its combustion laser.

As it turned out, the nozzle, which helped TRW win the DARPA contract, never worked, so researchers had to revert to using the original Aerospace nozzle. But in 11 months and under budget for the $10 million project, TRW produced a 100-kilowatt device that lased continuously for 60 seconds, a world record power.

In three short years researchers had progressed from lasing 1 kilowatt to over 100 kilowatts, and a weapons-class megawatt laser seemed just around the corner.

The DARPA BDL caught the attention of the navy, which moved to fund its own laser (NACL) for the navy–ARPA chemical laser. NACL was built at the Capistrano facility with a goal of producing a laser four times as powerful as the BDL.

By the mid-1970s, NACL had achieved powers on the order of 2.5 times that of the BDL, approaching 250 kilowatts. Although disappointing in the sense that it did not reach the original goal of increasing the power by a factor of 4, the navy's NACL set a world record in laser power.

With these increases in energy came other problems associated with beam quality, pointing and tracking, and even the effects of high-power lasers on the mirrors. Increasing the laser power no longer remained the long pole in the tent; other, more challenging problems began to surface.

Until now, standard reflecting mirrors could deflect a beam's energy without too many problems. As long as the beam was not focused on too small a spot, the beam would happily reflect off the series of mirrors in an optical train and be directed on its merry way to a target.

Now, however, with beam energies exceeding the bonding strength of the molecules that make up even a highly reflective mirror, merely exposing a reflective surface to such a high-intensity beam would cause the mirror to vaporize.

In addition, as the beam left the enclosure through an optical window—usually a piece of glass or similar transparent material that kept dust and debris away from the optical train—and as the laser grew in intensity, the optical window itself would be subjected to a large temperature rise and would either burn away or explode.

So as we've seen before, increasing a laser's power level was only one part of developing a weapons-class weapons system. The remainder of the components—high-temperature mirrors, high-quality optical trains, efficient optical windows, and acquisition, targeting, and tracking technology—was just as important as the laser if this dream of producing a speed of light weapon was to ever become a reality.

Acknowledging this discovery, the Unified Navy Field Test Program (UNFTP) made investments in acquisition and tracking technology, as well as using aerodynamic windows (windows made of fast-flowing air) to overcome some of these major system problems.

As the nation had moved away from the Apollo program to other high-technology endeavors, the HEL researchers involved in those seminal activities reads like a Who's Who in national security: Mal O'Neil (a Princeton Ph.D. who later as a three-star general commanded the Strategic Defense Initiative and is now the chief technology officer of Lockheed-Martin), Ed Gerry at DARPA, Bob Greensberg at Avco Space Systems, the navy's Walt Sooy, Captain Al Skolnick, Commander Pete Nanos (a Princeton Ph.D. who after serving as a vice admiral was tapped as director of the Los Alamos National Laboratory), Jack Martinez at TRW, and the air force's Don Lamberson (as a two-star general later headed air force S&T), Colonel (Dr.) John Rich, and Colonel (Dr.) John Otten, to name a few.

Seeing the success of the navy and air force lasers, as well as those produced under DARPA funding, the army decided that it wanted a battlefield-friendly laser for its own use—one that didn't take a cadre of Ph.D.s to run, didn't use an impossibly long logistics trail to supply, or didn't produce soldier-killing exhaust fumes.

The army's MAOS (Modular Array Demonstration System) eventually proved that an HF laser could have a fast start-up time for engaging enemy targets, just as the air force was producing the Airborne Laser Laboratory (ALL), a key technology demonstrator flown in the military's KC-135, a tanker version of the Boeing 707 aircraft.

The army originally wanted its lasers deployed in canisters, so that they would break apart in midflight and zap any obstacles in its path, such as mortars, artillery shells, or unmanned missiles. In addition, the army wanted a system that would absorb any chemical exhaust to keep troops safe. In other words, it wanted speed of light weapons to be a revolution in military affairs.

As laughable as these requirements in the mid-1970s might seem today (but *not* as laughable as the popular music of the day, such as "Disco Duck"), it is ironic that thirty years later, the army is about to deploy one of the first laser weapon systems possessing those same capabilities: MTHEL (mobile tactical high-energy laser), which is transportable, keeps its toxic fumes to a minimum, and has already demonstrated that it can kill artillery shells, unmanned missiles (Katyusha rockets), and even mortars in flight. So the moral of the story is that the vision announced by prophets of the revolution may eventually prove true.

In 1983 the military emphasis on lasers switched from tactical applications back to strategic defense with the establishment of the Strategic Defense Initiative (SDI). Although using directed energy to destroy ballistic missiles was only a small part of the SDI program, the effort was quickly dubbed Star Wars by a group of nontechnical skeptics, which some purported was politically motivated to keep both the ABM (antiballistic missile) treaty and the doctrine of MAD (mutual assured destruction) alive.[21] In reality, most proposed SDI systems needed advances in technology and may or may not have been successful.

To accelerate the technical maturity of proposed systems, Lieutenant General James A. Abrahamson, SDI's first director, simultaneously funded separate contractors to pursue different approaches in a "horse race" to achieve SDI's milestones. Several high-energy laser programs were started, including accelerating a highly promising discovery made in 1977 called a free electron laser (FEL, discussed later in this book).

In the mid-1990s, SDI was renamed BMDO (Ballistic Missile Defense Organization). Political pressure from the Clinton administration caused it to shift its focus from strategic to tactical defense. BMDO deemphasized long-term research in directed energy and concentrated on using kinetic energy kill vehicles (i.e., fast-moving, ground-based missiles such as the Patriot) to destroy incoming tactical warheads.

In the late 1990s, renewed interest grew in national missile defense as the Republican-controlled Congress pushed plans for a space-based laser (SBL) and the airborne laser (ABL). The services once again grew interested in the tactical applications of lasers, and collaborative programs such as the army's THEL (tactical high-energy laser, built by Northrop Grumman) with the Israeli government, and the special operations COIL-based advanced tactical laser (ATL, built by the Boeing Corporation), an advanced concepts technology demonstration (ACTD), were funded.

In 2002, BMDO was renamed the Missile Defense Agency (MDA) by the Bush administration, was reorganized, and dropped the space-based laser effort. In addition, all service programs such as the airborne laser were pulled under MDA to provide a single integrated missile defense program. There were some whispers in the Pentagon that the ABL move to MDA was partially motivated to keep the air force from using ABL as a slush fund for the F-22 program, the air force's number one acquisition priority. Regardless of the true motivation, however, once under MDA the nation's entire missile defense portfolio could finally be managed as a whole.

In 2005, the military's portfolio of high-energy laser weapons was funded at over $500 million a year, with most of the money going to ABL. The remainder was distributed to the joint U.S. army–Israeli mobile THEL (MTHEL), the special operations command ATL, the navy's FEL science and technology program for shipboard defense, and the army's laser large ordinance neutralization system (ZEUS).

The separate components needed for building a laser weapon—power, beam control, acquisition, and targeting and tracking—had all reached technical maturity. Now all the parts had to be integrated into a true laser weapon system, a formidable challenge.

# 7

# The Department
# of Defense Wakes Up

---

$A$T THE END OF THE 1990S, over $5 billion had been spent—primarily by the air force in the preceding three decades—on high-energy laser (HEL) science and technology.[1] During this time, researchers increased the power levels from milliwatt to megawatt in several types of lasers. This was an increase in over nine orders of magnitude—a factor of a billion.[2] Other advances were made in the supporting technologies needed to field a laser weapon system, such as beam control, adaptive optics, active mirror cooling, and modeling and simulation.

The separate parts were ready to be integrated into a true system. But the services had been working independently of one another, and each proposed laser weapon was at a different level of maturity.

There was no integrated plan for the military services to leverage advances made in the separate programs. To many, it was an uncoordinated clamor. With the invention of the laser, the services undertook a mad scramble to produce a weapon; often the research was performed in "stovepipes," with the different service programs persuing wildly disparate technologies to achieve their goals.

For example, through the Airborne Laser Laboratory (ALL) program, the air force had mounted an aggressive campaign in the 1970s and 1980s to demonstrate that a laser on an airplane could shoot down air-to-air missiles. Using the lessons learned from the ALL, the airborne laser program was established as a systems program office at the air force Phillips Laboratory in 1994 to build the ABL. After leveraging the Phillips Lab's technical expertise to lay the groundwork for the technical requirements, the ABL SPO was moved directly under the air force Space and Missile Center; in October 2001 it was moved into the nation's integrated missile defense plan under the Missile Defense Agency.

In parallel to the air force effort, TRW's baseline demonstration laser (BDL) was the first chemical laser to lase with more than 100 kilowatts of power in the 1970s. The BDL follow-on, NACL (for navy-ARPA chemical laser), housed at the TRW San Juan Capistrano facility north of San Diego, achieved its first light in 1975.

The navy's effort to demonstrate a large ground-based laser culminated with the MIRACL (mid-infrared advanced chemical laser). Located at the White Sands Missile Range High-Energy Laser System Test Facility (HELSTF), MIRACL is a megawatt-class combustion-driven DF chemical laser that first lased in 1980. Built for the Sealite program, it was the Defense Department's most powerful laser until the ABL fired.

HELSTF also was home to a smaller laser developed by the air force called EMRLD (excimer midpower raman-shifted laser device). The air force shut down research on EMRLD on October 1, 1989.

Finally, DARPA initiated the Alpha laser program with the goal of developing a megawatt-level space-based laser demonstrator that was scaleable to more powerful weapons levels and optimized for space operation. Alpha achieved megawatt power levels in 1991 similar to

MIRACL, but in a low-pressure, spacelike environment, demonstrating that weapons-class, space-compatible lasers were possible.[3]

Recognizing the need to pull together these disparate military research efforts, Dr. Delores Etter, then deputy director for defense research and engineering (DDR&E) in the Pentagon, chartered a group of government experts in September 1999 to focus the nation's investment in high-energy lasers, and specifically for use as weapons. The Department of Defense recognized that high-energy lasers were going to play a key role in future weapon systems, and the nation was best served by consolidating and leveraging the disparate research programs conducted in the military services.

The study group focused in six key technology areas: chemical lasers, solid-state lasers, free electron lasers, beam control, lethality, and advanced technology. With the help and advice of outside experts, the committee reached six conclusions, which serve as the foundation for today's directed energy master plan:

*1. High energy laser systems are ready for some of today's most challenging weapons applications, both offensive and defensive*

Enough progress had been made over the past four decades in beam quality, control, and power that the committee recommended the Pentagon demonstrate the military utility and effectiveness of HEL systems. The committee recognized that although significant engineering obstacles remain to deploying HEL systems, there is no overriding physics-based limitation to building air, ground, and sea-based HEL weapons.

The committee based its conclusion on the fact that disparate parts of a HEL weapon system had been demonstrated—megawatt power levels by Alpha and MIRACL; beam quality enhancement through adaptive optics by the air force's Starfire Optical Range; advanced beam control and steering by the navy's Sealite beam director; and laser lethality and system vulnerability experiments and analysis by all the services. However, the parts had been built separately and had not yet been integrated into a single working system that would withstand the rigors of the battlefield.

As expected, this recommendation created debate in the scientific community. On one side were the people who saw this as just a hard engineering problem and advocated optimizing the laser technology and building a HEL weapon by integrating the disparate parts. Since no basic laws of physics were being broken and the major components of a HEL system had been demonstrated years before, those advocating this position were puzzled at the uproar.

In contrast, those who were horrified by this philosophy cited the inherent difficulties in being blinded by optimism undertaking such a high-risk endeavor. After all, wars are fought on battlefields full of smoke, uncertainty, danger—the constantly evolving "fog of war"—and certainly not in a pristine laboratory where hundreds of Ph.D. scientists are needed to tweak a HEL system to make it work. Simultaneously increasing the power of a HEL, controlling the beam, making the beam quality high, and keeping it stable and pointed on a target—in a battlefield environment—is more than just an engineering problem. It's a major undertaking fraught with risk.

The late Dr. Robert Forward pointed out both the fallacy and the seduction of understanding this challenge as just an engineering solution. Just because nothing says that it can't be done, doesn't mean that it *can* be done.

Got that?

For example, it breaks no law of physics to create zero gravity. Most people are startled by this revelation, but it's true—ask any physicist. There is no physical reason why you can't create a volume of zero gravity on earth.

Albert Einstein used to conduct *gedanken* experiments to examine difficult concepts, so let's consider one that creates zero gravity:

Think of a place in deep interstellar space, far away from any object, so that gravitational effects are very small. Since gravity is caused by mass, even the presence of a small amount of mass—such as a base-ball—creates gravity.

An interesting thing about gravity is that it is known as a vector field. In simple terms, this means that gravity has both magnitude and direction. It's the reason why you fall at a certain rate (magnitude) and down (direction) when you fall off a stool.

In our *gedanken* experiment, the baseball creates a small gravitational field that has both a magnitude and a direction. Objects placed in this gravitational field are accelerated toward the baseball with a magnitude (rate) and a direction (toward the baseball).

Now suppose we have two baseballs in our experiment. It shouldn't be surprising that each baseball creates its own gravitational field. But here's the amazing part: there is a location directly in between the two baseballs where the gravitational field is *zero*; the two gravity fields cancel each other out and as a result there is a small volume of zero gravity.

This is because of gravity's vector nature: although the two fields have the same magnitude, they are in opposite directions to each other, and the directions cancel out. We see this phenomenon in our solar system at the so-called Lagrange points where the sun's, earth's, and moon's gravity cancel one another out.

So we can see it's possible to create zero gravity. You can do it on earth. It's even possible to have a volume of zero gravity the volume of a football stadium. Theoretically, that is, since it is "just an engineering problem."

To create a volume of zero gravity on earth the size of a football stadium, you'd need a mass the size of the moon squashed to a slab a few meters high to cancel out the direction of earth's gravity. No problem, except this mass would have to be supported a few hundred feet off the ground.

In other words, this would be an impossible engineering problem, even when there is no physical reason why it can't be done.

That's not to say that building a HEL weapon system is as hard as cre-ating a volume of zero gravity on earth! But the point is that just because there is no reason why a complicated system won't work, is not to say that it *will* work.

### 2. HEL weapons offer the potential to maintain an asymmetric technological edge over adversaries for the foreseeable future

If diplomacy fails and no option remains except a call to arms, no one wants to return to the old way of fighting wars, slugging it out toe-to-toe against an evenly matched adversary. War is bad enough without being a war of attrition. As national leaders have said since the first Gulf War, we don't want a fair fight—war isn't a basketball game; rather, it should be accomplished as swiftly as possible with overwhelming force and advantage.[4]

In the past, nations that exploited new technologies in warfare gained the upper hand. An asymmetric advantage is a technique of turning the tide of battle without matching a foe head-to-head.

At the air force centennial celebration of flight, a paper on leveraging directed energy in asymmetric warfare recalled these historical facts:

In ancient Greece the phalanx—a close-order fighting formation of heavily armed infantry spearmen troops—with its discipline and speed of maneuver allowed Greek warriors to dominate their part of the world for centuries. Roman Legions—a coherent army combat unit of foot soldiers and cavalry capable of speed and agility—used the Roman short sword as a thrusting weapon at close range, which provided Rome the basis to conquer the world as it was then known. Introduction of the stirrup in the sixth century allowed a horse and rider to go from a means of transportation to a lethal shock weapon combining the mass and speed of the horse with the thrust of a lance or saber. The armor-mounted knight of Europe was neutralized by the stand off firepower of the long bows of the English yeomen. Over time, long bows and massive fortifications were neutralized by the invention of gunpowder and siege cannons. During the age of discovery, empires were built on the range and firepower of sailing ships only to be superseded by ships powered by steam. With the invention of rifled barrels and rapid-fire weapons land warfare later came to be dominated by trench warfare.

In the twentieth century trench warfare was replaced by maneuver with the adaptation of the internal combustion engine to machines of war. Aircraft emerged to provide the ultimate high ground combined with speed, range and lethality. Later in the century missiles made it possible to strike any place on the globe within minutes. This same technology made it possible to lift highly effective reconnaissance and communications payloads into space. Combined with air breathing intelligence, surveillance, and reconnaissance assets, these space assets began to pave the way for a new era in warfare, one in which it would be possible to find, fix, track, target and engage anything of consequence that moved on the face of the earth or through the atmosphere.[5]

Etter's committee used this rationale to illustrate directed energy's asymmetric advantage over traditional weapons systems. The main reason deterrence worked for 50 years in the Cold War, and the main reason the Soviet Union felt it could not go toe-to-toe against the United States was the existence of America's nuclear arsenal. For the Soviet Union (as well as the United States), fighting a traditional war against an enemy that had the ability to obliterate you didn't make sense.

Today, times have changed. Terrorists are no longer confined to single nation-states; they move undeterred across the world, and plan their activities to asymmetrically strike against our most cherished possessions, our center of mass. Going after terrorists in a populated city with bombs and bullets will only cause them to hunker down and wait it out; but directed energy weapons, such as Active Denial or ATL, provide an asymmetric option to respond to their threat.

A rogue nation might spend billions trying to perfect an intercontinental ballistic missile and might use that capability to threaten its neighbors or the world.

But in the same way that Active Denial might provide an asymmetric advantage in urban warfare, the airborne laser could provide an assymmetric advantage by negating a country's ICBM capability by destroying a missile in flight. Asymmetric response: the Department of Defense believes that directed energy is poised to provide this solution for years to come.

### 3. Funding for HEL science and technology (S&T) programs should be increased to support priority acquisition programs and to develop new technologies for future applications

This recommendation is self-explanatory. One would think that the Department of Defense would prioritize its programs and, with the money available, move down the list and fund what it can. After all, that's what everyone does with a household budget and commercial companies do it to stay afloat.

But realistically the Defense Department cannot be run like a home budget or a company on a cash-flow basis for a couple of reasons. The congressional authorization and appropriation process is contorted, and the government's product is national defense.

And that involves making hundreds of thousands of acquisition decisions. The government's priorities range from funding chemical and biodefense to building training facilities. These are all important, so who's to say where directed energy ranks in this megapecking order? The acquisitions process in Washington has been likened to riding through whitewater rapids on a leaking budgetary raft.

So how does the government accomplish this recommendation from the directed energy panel? Especially with the disparate priorities, and particularly with the different service programs in high-energy lasers, all competing for essentially the same pot of shrinking science and technology (S&T) dollars?

One way to do this is to establish a single point of contact responsible for overseeing the health and future of S&T investments for HELs.

The Joint Technology Office (JTO), discussed below, was established in 2000 to do exactly this. Initially led by Dr. George Ullrich, a physicist from Dr. Etter's office, the JTO was then directed by Ed Pogue, a Los

Alamos scientist who currently oversees investments on the order of $75 million a year to support the nation's HEL infrastructure.

*4. The HEL industrial supplier base is fragile in several of the critical HEL technologies and lacks an adequate incentive to make the large investments required to support current and anticipated Defense Department needs*

High-energy laser systems require a support base unlike any other in the world. An air force four-star general once asked laser researchers why this is true. After all, he reasoned, if the telecommunications industry supports billions of dollars in research, why can't the military leverage its spin-offs? Out of the legions of companies jockeying to dominate market share in telecommunications, entertainment, or even medical lasers, why can't advances they have made be applied to laser weapons?[6]

The answer lies in the difference in the magnitude of power levels. Strategic laser weapons are a billion times more powerful than most commercial lasers, and that makes all the difference in the infrastructure needed to support this technology—from mirror coatings to alignment to pointing-and-tracking requirements, and to the storage and safety of the laser fuel.

There is no demand in the telecommunications industry for materials that can withstand the heat generated by megawatts of power because even the highest-power lasers used for transmitting telephone calls, digitized Internet protocol packets, and video streams are a hundred thousand to a million times smaller in power than what the airborne laser will use.

Consider the following. The entertainment industry uses low-power solid-state lasers for CDs, DVDs, and laser light shows. It can't afford to injure people with devices seemingly as benign as laser light pointers. The medical field needs low-power energy in applications that range from laser eye surgery to dermatological smoothing and tattoo removal.

Low-power laser communications can provide higher bandwidths, or the ability to carry more channels, than standard microwave transmission systems, and those laser systems need to be accurately aligned as lasers project a much smaller "footprint" (or area) than microwaves. But these laser communication systems are not tracking ballistic missiles or Katyusha rockets, and they are not being transmitted through hundreds of kilometers of turbulent air. They do not need either the highly accurate pointing-and-tracking accuracy or the atmospheric compensation gained by installing deformable mirror technology.

In addition, the telecommunications industry does not have to worry about the logistics trail or the health hazards associated with the fuels used for chemical lasers, as the low-power solid-state lasers used in fiber optics are engineered to be safe and easy to use.

As a result, since the commercial laser industry is driven by market forces and those forces are focused on maximizing profit, there is little motivation for industry to invest in the sometimes finicky (and somewhat whimsical) government-dominated high-energy laser market.

Industrial support gravitates to the largest market share, and high-energy laser weapon systems are such a small part that HELs have lost nearly all influence in the market. After all, why should industry support a niche that is less than 1 percent of the total market—especially when long-term investment in HEL science and technology is dictated by government needs, which fluctuate like dust in the wind?

In the 1960s and 1970s, the government drove the development of directed energy by making large investments in laser science and technology. Every military service conducted HEL research, and the government pumped several hundred million dollars a year into the economy to enable this research.

At the same time there was growing commercial application of lasers, but not in high-energy lasers. Heavy industry, such as precision metal drilling, cutting, and tooling, were the only widespread users of commercial HEL technology, but this remained a small fraction of the total.

In the 1980s, when low-power laser applications in fiber optics, medical surgery, light shows, and the birth of CD and DVD technologies started to increase, the number of commercial uses for lasers soon surpassed that of high-power military lasers. And near the end of the SDI heyday, when HEL funding plummeted, commercial investment in lasers dwarfed that being spent by the military.

Suddenly the Defense Department was neither dictating the direction of laser research nor driving the laser train. The military could no longer dangle money in front of industry and have companies scramble to invest their own IR&D (internal research and development) funds to advance military laser science and technology. And if the military could not leverage research that was already occurring in the commercial sector, it would have to pay dearly for what it needed.

And those funds were being eyed by other factions of the military that were competing to fund their own programs, such as the F-22 fighter, the army's future combat system, or the navy's new carriers and submarines.[7]

The phenomenon of losing influence when moving from a government-dominated sector to a commercial-dominated sector is not new. Since the late 1940s, supercomputers built for the government solved national security problems ranging from breaking crypto codes to designing nuclear weapons. But as the supercomputing field matured and first university, then commercial researchers discovered their own non–national security applications, the government share of these boutique computing machines dwindled.

Today the National Science Foundation high-performance computing centers (HPCC), university and industrial HPCC, as well as foreign advances in supercomputing (such as Japan's numerical earth simulator, which in the early twenty-first century was the fastest computer in the world) dominate the supercomputing market.

With few exceptions, the government no longer dictates the direction of supercomputer technology. Nearly all aspects of high-performance computing are now influenced by market forces.

In this environment of having to work at the margin in laser research and development, Etter's committee realized the military needed a way to focus what little leverage it had to advancing critical HEL science and technology—and not play the politically expedient game of spreading money as wide as possible, which gives something to everyone but accomplishes little.

### 5. The Defense Department should leverage HEL-relevant research being supported by the Department of Energy (DOE) and other government agencies and also by commercial industry and academia

There's a saying that made the rounds a few years back when the dean of science fiction, Robert A. Heinlein, was still alive: "TANSTAAFL." There ain't no such thing as a free lunch. In other words, you don't get something for nothing.

In a perfect world you would expect that our government—you know, the one you pay taxes to support—would optimize its resources and ensure that one part of the government helped the other part.

For example, Lawrence Livermore National Laboratory is known as the Department of Energy's laser laboratory. Doesn't it therefore follow that research performed there would be used in various parts of the government? After all, Livermore is a national lab, not a commercial company accountable to its stockholders to produce a profit.

In a sense, the entire government does get to tap into the benefits generated by Livermore's research. Young scientists are trained, technicians gain experience, and advances are made in certain areas of laser physics.

But the truth is that the Department of Energy, which runs Livermore, has a mandate to produce fusion with high-energy lasers (embodied by Livermore's National Ignition Facility—NIF), not to produce HEL weapons.

Although lasers for fusion and lasers for weapons are both lasers, comparing the two is like saying a Boeing 777 passenger jet is similar to an air force C-17 cargo jet. They're both jet airplanes, and they both carry immense amounts of weight. But the similarity stops there. The C-17 can take off and land on 3,000-foot dirt runways; it can fly arduous, low-level, stress-producing missions that strain the limits of mechanical integrity; it can refuel in the air; and it can carry tanks and troops to a combat zone with a crew of three.[8] On the other hand, the 777 can be configured to ferry up to 550 passengers, has demonstrated an unrefueled range of over 20,000 kilometers, burns fuel at a rate less than any other jumbo jet, and sets the standard for commercial airliner efficiency.

In short, although the C-17 and 777 are both state-of-the-art jets, their applications are orthogonal. But this comparison may be too close, when using an analogy of lasers for weapons and lasers for fusion. Comparing military jet fighters with the 777 is even better in this respect, but also is more stark.

Fighters fly for short, intense periods, sometimes fractions of an hour—usually in combat, where their engines and airframes are stressed by G forces spanning over 7 Gs. In contrast, a 777 can fly for over 21 hours at a time, and there are few inflight changes in altitude, speed, or direction. There are many differences between the two different platforms.

Similarly, laser research at Livermore may yield fundamental advances for military HEL research. But why should the Department of Energy use its scarce dollars to boost a military mission? Or conversely, why would the military—charged with defending our nation—use its own scarce dollars to increase the probability that a laser could achieve fusion?

The answer is that at the highest level, they both work for the same government and they should do it. But the reality is that in these times of constrained budgets, competing priorities, demanding schedules, and intense competition for resources, the probability of obtaining leverage from disparate agencies is vanishingly small.

### 6. It is increasingly difficult to attract and retain people with the critical skills needed for HEL technology development

This observation is true not only for industries such as HELs, with scarce skills in laser physics, optical coating, sensor development, and computer science, but for any high-tech industry, and especially for those focused on national security.

The number of foreign students attending U.S. graduate schools for degrees in the physical sciences skyrocketed in the 1990s and leveled out after 2000. Simultaneously, the number of U.S.-born students has declined, meaning that there are fewer graduates with the proper education and background needed for security clearances to perform high-tech research in national security fields, such as high-energy lasers.

Also driving this scarcity of resources is the escalating demands of the commercial industry. Annual commercial investment in the information and electronics industry rose over 20 percent in the last four years of the 1990s, reaching $77.8 billion in 2000, overwhelming the defense science and technology investments in these areas.[9]

As a result, this commercial demand for high-tech talent, coupled with the lower pay available to government researchers, means fewer workers will gravitate to high-energy lasers. This analysis does not take into account even worse trends, such as the dwindling numbers of students who are enrolling in science and technology fields as compared to past years. Thus it is no surprise that there is a dearth of talent entering the fields associated with HELs.

In response to this problem the commercial industry "banks" talent. That is, because commercial companies have a bottom line of profit, they are generally reluctant to hire more talent than they presently need—witness the massive layoffs by commercial aerospace companies if a large plane order is canceled; California aerospace companies were flooded with engineers after the end of the Apollo program (and recall that TRW in the late 1960s moved into HELs after identifying this area as the "next big thing" after Apollo).

Banking talent means hiring bright, motivated professionals in anticipation of the "next big thing," and this philosophy has enabled some high-tech firms to load-level their hiring of scarce talent for when the next demand hits their company.

But this practice is not without risk. Keeping talent around, no matter how promising, is a drain on the company's bottom line: profits. And if the company is infused with the so-called Harvard MBA mentality of only looking at the next two quarters of company profit, then that entity will go the way of other technological giants—such as Bell Labs, which once had more Nobel Prize winners per capita than any other laboratory in the world. Finally, however, it found itself in a death spiral because of ever-increasing demands to pay homage to the bottom line.

Etter's main conclusion was that the Defense Department should target certain S&T areas (such as optical fibers or optical coatings for mirrors that can reflect high-energy lasers) and focus its investment in those critical, high-leverage technologies. But that would take an entirely new entity focused on ensuring the health of high-energy lasers.

And that's just what happened. The government chose to step in and seed various critical S&T investments in universities, industry, the military services to ensure that the necessary infrastructure exists for the development of HELs.

As a result, the Joint Technology Office was set up in 2000 to "advocate and execute" the S&T investment for high-energy lasers, and it currently invests on the order of $75 million a year to ensure this mission.[10]

# 8

# Prophets of the Revolution:
# High-Power Microwaves

THE SUMMER OF LOVE: 1969. It was a pivotal year in many ways, and we may never see another one like it. It saw the first man on the moon. War. Peace. Student protests.The Beatles. Drugs. Social disruption.

The promise of lasers seemed to rule the future and no one thought there might be any other path for national security. It was only a matter of time before the directed energy age would be acknowledged as the successor to the atomic age, especially now that Avco Space Systems had broken the 100-kilowatt limit—the DE equivalent to the four-minute mile—and megawatts of laser power seemed to be just around the corner.

The era brings back a flood of memories for those who experienced it and for the young, tales of disbelief as they read about clashing cultures, the end of a turbulent decade fueled by the juxtaposition of the promise of the space age and the in-your-face protests of rock and roll. It was a new outlook on technology, on sex, drugs, and changing lifestyles, a lurch away from Ozzie and Harriet to Ozzie and Black Sabbath. In terms of culture and technology, it was an invigorating yet weird and unstable time to be alive.

So in June 1969, armed with a new Ph.D. in nuclear physics from Ohio State University, Bill Baker was ready to conquer the world. It seemed that science could solve any problem.

The laser had been invented nine years before, and with the United States on the verge of sending a manned mission to the moon, nothing seemed impossible. And now the air force wanted Baker to join other young, highly educated scientists in its quest to exploit science to win the Cold War. In Baker's mind, it didn't get any better than this.

As the United States was being turned upside down in a cultural revolution, the air force was assembling a great collection of scientists and engineers in a move not seen in years. Only the gathering of scientists at Los Alamos during World War II to develop the atomic bomb or NASA corralling the nation's aerospace expertise during its rush to mount a mission to the moon could come close to the breadth of science talent being assembled at the Air Force Weapons Laboratory.

The Vietnam-era draft in the 1960s and early 1970s steered young men to graduate school to postpone being sent to Vietnam. Some hoped that an advanced degree would land them a cushy job performing research in a defense lab instead of slogging through the jungles of Southeast Asia. Others saw a chance to serve their nation by doing what they did best—using their scientific talent to help win the Cold War. In both cases, the services were flooded with highly trained officers holding Ph.D.s from powerhouses such as Princeton, MIT, Yale, Columbia, Cal-Tech, Michigan, and many others.

The air force opened its arms to these highly educated young officers, and many were sent to the Air Force Special Weapons Center. The center was established after World War II to study the effects of nuclear weapons on aircraft, electronics, and other military systems. It was also the home of the fledging air force high-power laser program, soon renamed the Air Force Weapons Laboratory (AFWL), reflecting hope that lasers would dominate the exotic weaponry of the future.

But when First Lieutenant Bill Baker arrived at the desert research station located at the far end of Kirtland Air Force Base (AFB), the high-tech facility looked like a slum. Small, squat buildings covered in brown stucco with traditional Mexican-style flat roofs dotted the deserted west side of the base. Set at the far southeastern side of Albuquerque, Kirtland AFB jutted up against the Monzano mountains, defining the corner of Albuquerque. The wind-whipped base was brown, flat, and arid.

As he looked over the collection of run-down buildings, Baker felt his initial excitement at being assigned to the AFWL turning into despair. He had just left a cultivated land of green trees, rolling hills, lakes, and midwestern sensibility. Here the hot desert air swirled dust around buildings.

Like the other officer-scientists before him, Baker was given the opportunity to interview through the Air Force Weapons Laboratory to cast his lot where he would like to work. At the time the AFWL concentrated in two main areas of research—lasers and nuclear technology.

The fledgling laser area was headed up by Don Lamberson, a gruff young major who held an engineering Ph.D. from the Air Force Institute of Technology in Dayton, Ohio.

Lasers were sexy, as directed energy was then the darling of the air force. It captured the air force fighter pilot image—high-tech, cutting edge, and futuristic enough to convince the generals that it was a trump card for winning the Cold War. In fact, a highly classified study called Eighth Card—so named as how one could win seven-card stud with an illegal trumping eighth card—determined that battlefield lasers could defeat an adversary who otherwise matched the United States talent for talent; in other words, provide an asymmetric advantage.

Lasers were the future, and the eyes of the air force were riveted on them. Researchers flocked to join the exciting team, as the effort was being led by the dynamic, no-nonsense Lamberson.

In contrast, the nuclear technology group researched the largely unknown field of nuclear weapon effects, investigating phenomena such as airblast, cloudrise, ground shock, and X ray damage. Nuclear effects was a core air force research area, but it was not as flashy as lasers.

In addition, Lamberson's take-charge, take-no-prisoners personality contrasted with Baker's, who was low-key and introspective. Lamberson had surrounded himself with a cadre of officers like himself—aggressive, quick on their feet, opinionated—a world of difference from the thoughtful, intellectual university culture Baker had just left.

Despite the optimism about lasers, not everyone was enthralled by their promise. It was easy to project engineering advances on a vu-graph slide and forecast technological breakthroughs in front of military generals who were searching for any technological advantage over the Soviet empire.

But Baker didn't buy into it. He was fascinated by the research being performed in nuclear effects and in particular studying the generation of high-energy X rays, similar to those that would be produced in a nuclear explosion.

Electing to throw his hat in with this crowd, Baker joined a small circle of intellectuals starting to ask, "What's Plan B?" In other words, what happens if lasers don't work as advertised?

Little did Baker know that in a few years he would be at the center of a scientific breakthrough that would herald the coming of an unforeseen epiphany, a flash of inspiration in the guise of high-power microwaves: Plan B.

High-power microwaves, now simply know known as HPM, had been a dream of the military for years, ever since electronic warfare proved that radars could be jammed. For in the late 1950s, it was discovered that high-power transmitters, finely tuned to enemy radar, could overpower Soviet antiaircraft radar being employed against U.S. nuclear bombers.

SAC—Strategic Air Command—controlled the hundreds of nuclear-carrying B-47s and B-52s that countered the perceived Soviet threat. Enemy radars threatened to defeat SAC's mission to defend

America, virtually negating billions of dollars that had been poured into the massive nuclear air strike system. But now electronic warfare, in the guise of low-frequency microwaves, promised to halt the Soviet countermeasure.

The Soviets countered America's electronic warfare threat by producing radars that shifted frequencies. They "hopped" from band to band like a football player jinking from side to side as he runs down the gridiron in order to keep from being tackled.

American researchers were frustrated. They tried to raise the power of their jammers and even make them smarter by anticipating the Soviet jump in frequency, but nothing seemed to work. The Soviets encased their radars in electromagnetically hardened shells called Faraday cages, after the English scientist from whose equations showed that it was possible to electronically shield equipment from electromagnetic radiation. Although still a formidable weapon against less sophisticated radars, especially those fielded by third world countries, electronic jammers seemed to have met their match. It appeared U.S. forces could not defeat Soviet radar.

But during this high-tech point and counterpoint of electronic jamming, it was realized that some kinds of microwaves could slip around the hardened metal Faraday cages that enclosed delicate electronics and wiggle into the circuits that commanded the sensors. The microwaves could do this without being tuned to the enemy's transmitting frequency, something the enemy could successfully jam by overpowering our countermeasures.

This was a startling revelation. Instead of going toe-to-toe against the enemy radar by attempting to overpower or "spoof" their signals— by going through the "front door"—or even by trying to follow their sophisticated frequency-hopping algorithms, this approach used an entirely different technique: going through a "backdoor" to slip inside the electronics. This technique had been know for over 30 years but had never been tried because microwaves did not yet have enough power.

The image of going through the "backdoor" evokes a stealthy character slipping around the side of a house during a well-attended party. While the hosts may have posted guards to watch for anyone trying to crash through the front, our stealthy intruder squeezes through a small window in back, slithering into the house undetected to create mayhem.

This is precisely the secret of high-power microwaves.

Normal electronic warfare—known as jamming or spoofing—relies on a quick determination of the frequency that the enemy is broadcasting. With advances in solid-state circuitry, this can be done in a few thousandths of a second. A computer can analyze—again, in a few thousandths of a second—how to fool or spoof the enemy. A fraudulent signal can then be broadcast that dupes the enemy and thus jams his radar. Another brute force jamming technique is to overpower the enemy's transmitter by broadcasting so much energy that enemy circuits are fried.

Both jamming techniques can be countered by using filters, the electromagnetic equivalent of installing fuses to prevent the transmitters from being overpowered. These methods of electronic warfare are all known as going through the front door—trying to defeat the enemy's transmitters by slugging it out head-to-head, toe-to-toe, each trying to outwit the other.

This point–counterpoint approach to electronic warfare results in an ever escalating complexity of feint and counterfeint. In many ways, it's a technological death spiral, with the latest inventor prevailing.

However, going through the backdoor is an entirely different way of jamming. And doing it with HPM is like conducting electronic warfare on steroids.

Instead of trying to defeat an enemy by making minuscule improvements in electronic warfare that you hope the enemy won't notice, using HPM is like kicking down the door and throwing in a hand grenade. HPM doesn't care about the front door. In some cases, it's like the front door doesn't even exist. Take no prisoners.

At its highest power levels, HPM can blast through any filter and overpower the electronics to make them a pulpy, metallic mess. This is the high end of the scale, the stuff that Hollywood loves, *Terminator* effects, where the electronics are fried and the control room is smoking.

At the opposite, lower-power end of the spectrum, microwaves can slip around an electronic device and slither through cracks, seams, and even transmitting wires to gain entrance into the inner sanctum of an enemy's weapon—the computer, or the brains of the device itself.

Once present, the microwaves can subtly change the very nature of the electronic computations and make the sophisticated circuitry "dumb." Since no one can perfectly shield a weapon from these HPMs, microwaves have the ability to cause damage in a full-spectrum manner, from denying the enemy access to its weapon to destroying it.

This is profound.

Up to that point, we only had the ability to electronically jam a target. By slowly raising the power of microwaves that are inserted into the target, researchers now have the ability to create a full spectrum of havoc, from deceive, to deny, to damage, to destroy—or as some scientists say, "from toast to roast."

And the best part of it all, the "secret" of high-power microwaves, is that this damage can be done through the backdoor, where there is no defense or way to stop it.

Prior to this discovery, hundreds of millions of dollars had been poured into systems that tried to defeat newly emerging enemy electronic weapons.

With lasers, military leaders envisioned transmitting powerful beams that would precisely knock missiles out of the sky. But even in the early years of laser research, it was realized that an enormous amount of energy would be needed to make a useful laser weapon. Generating weapons-level HPM was much harder to achieve, so much so that many people thought that it would be impossible to ever achieve.

On the other hand, researchers had been generating microwaves for decades.

First used in high-power radars, microwave tubes produced the high power levels needed to detect enemy planes miles away from the border.

These earliest uses motivated some researchers to exploit microwaves as weapons. The Japanese investigated using HPM as weapons to down airplanes in World War II.[1] Their efforts failed because they were not able to obtain the intense fields needed to affect Allied flight controls.

Microwave tubes grew in utility as the communications and aerospace industries expanded their use. Powerful microwave tubes were developed for applications ranging from detecting incoming intercontinental missiles to accelerating exotic streams of particle beams at universities.

Throughout this development, it was a well-known fact that electronic equipment would stop working when brought next to TV and radio towers. Radios would stop working, electronic door openers in cars wouldn't function. At some locations, people would even hear clicking noises or feel nauseous. The one thing these radio frequency waves all had in common was that they could be classed as HPM. There appeared to be more to high-power microwaves than met the eye.

In the 1960s, the air force envisioned powerful microwaves blasting enemy aircraft out of the sky, disrupting their electronics just as Japanese scientists had predicted 20 years before. When air force researchers calculated the power needed to produce these effects, they realized that it would take years of development to produce enough power in microwave tubes to even come close to accomplishing this mission. As a result of this study, the air force refocused its research toward the more promising field of lasers.

In the 1970s the navy started investing in HPM research and even built 40 giant microwave tubes (klystrons) at a research facility in San Juan Capistrano, then a desolate research area north of San Diego.

But after millions of dollars had been spent on construction, the FAA refused to allow the navy facility to come online, as it had not been realized until then that the klystrons would radiate in a frequency close to that used by the Los Angeles air traffic control. It was a disaster waiting to happen, and the navy shut down the multimillion dollar facility before it ever became operational.

After the air force dropped its HPM research in the 1960s and the navy was fiscally whacked in the 1970s, nearly a decade passed before the army showed appreciable interest in this dubious field of HPM.

Researchers at the Aberdeen Proving Grounds in Maryland realized that the increasingly sophisticated electronics that the enemy—and worse, the United States—was fielding would be susceptible to electronic warfare. Much progress had been made in HPM sources, making the growing technology more and more attractive to war fighters. HPM was slowly moving from a scientist's fantasy to reality. By 1984 the army started investing in HPMs, adding fuel to the field.

This growing interest did not result from advances in the traditional path of microwave development—by incremental improvements made in vacuum tube technology—but rather because of quantum leaps in power made by ultra–high current particle beam physicists.

This demonstrates that sometimes a dramatic change in capability is made by an expert in one field working in an unrelated field. HPMs are such an example.

## The Lure of Particle Beams

Flashing back to 1973, Captain Bill Baker resigned from the air force and accepted a civil service job at Kirtland AFB as science adviser for the Air Force Weapons Lab's X Ray Simulation branch.

The new job was a comfortable position, and Baker was well qualified for it. The spit-shine military lifestyle of haircuts, frequent moves to desolate places, and unquestioned obedience didn't sit well with the introspective, ever inquisitive Baker. He'd been top of his class at Ohio, and having military officers not schooled in science micromanaging his work did not appeal to the talented young researcher.

Over the past four years Baker had made enormous contributions to his research group. While his contemporaries in Lamberson's laser program had garnered much attention from air force generals while demonstrating increasingly powerful laser pulses, Baker had worked in a less glamorous field. He wasn't content to follow the laser fad or simply be one of several hundred physicists needed to ensure the success of Lamberson's highly acclaimed Airborne Laser Laboratory (ALL), a demonstration of an airborne high-power $CO_2$ laser that would shoot down an air-to-air missile.[2]

Baker had pioneered far-reaching advances in implosion physics, an arcane field familiar mostly to American and Soviet nuclear weapons scientists. His group was trying to create X rays similar to those produced in a nuclear explosion so that the military could test its most sensitive equipment in a laboratory, rather than spend hundreds of millions of dollars conducting a brief (less than a thousandth of a second) nuclear test deep underground in the Nevada desert.

Baker's collaborators would include some of America's most influential physicists, such as Dr. Edward Teller, father of the H-bomb, as well as a virtual Who's Who of Soviet scientists. Names such as Mohkov, Lebedeev, and Panofsky, unknown to most Americans yet heroes in the Soviet Union, graced the ranks of the Russian academicians—the rough equivalent of the American National Academy of Sciences except much more influential, and held in esteem worthy of a rock star. Later these scientists were among the first Soviet citizens to visit America.

Baker's group was charged with investigating exotic weapons concepts, which gave birth to the Advanced Concepts branch. With the overattention that had been given to lasers, Baker was ready for a new challenge, something that could help national security yet be intellectually stimulating.

Urban myth holds that one of Baker's scientists, a wild-eyed Ph.D. from Montana, once showed Baker pictures taken from huge astronomical telescopes, images light-years away that showed jets of particle beams streaming in a line billions of miles long. Major Marv Alme, known in the advanced concepts cabal as the Montana Wild Man, had obtained a Ph.D. from "Teller Tech," the University of California–Davis campus located at Lawrence Livermore National Laboratory. Edward Teller had been instrumental in establishing Lawrence Livermore Laboratory and in founding the small graduate school that still unofficially bears his name.

Known for smoking cigars while hammering out code (sophisticated mathematical software), Alme made millions in rental property and housing renovation. In his day job, Alme enjoyed living on the technological edge, pushing the frontier in plasma physics, the esoteric subfield that made up over 99 percent of the universe.

Plasma is regarded as the fourth state of matter, as different from gas as liquid is from solid. Plasmas are a hot, soupy mixture of raw electrons and ions, "star stuff," the material that makes up stars and exists in the interior of a nuclear explosion. As science adviser, Baker rode herd over a host of eclectic plasma physicists hailing from Princeton to Teller Tech, and Alme was the chief theoretician.

Cigar dangling from his mouth, Alme collared Baker in the hall of Building 322, home of the Advanced Concepts group. He was holding a recent copy of a popular scientific journal. Alme stabbed a finger at the picture of the interstellar particle beam, a long, cigar-shaped, white-hot jet that emanated from the core of a stellar object. "If nature can shoot lightning billions of miles without flying apart, why can't we do it on a smaller scale?" Then Alme pushed past, having made his point.

Baker was struck. It seemed so obvious. Particle beams: man-made lightning. Baker was familiar with using charged particle beams to produce nuclear effects, but using them as man-made lightning weapons was a new twist.

Scientists had long been mesmerized by lightning. Lighting strikes continuously bombarded the earth with enormous amounts of energy, directing beams of highly energized particles—electrons and protons, the basic constituents of matter—hundreds of times a day. And not all lightning strikes occur from air to ground. Nearly half of the lightning strikes viewed by satellites propagate from clouds to the ionosphere, to distances far above the earth. These strikes are nature's attempt to neutralize its overall charge distribution by exchanging charged particles between the upper ionosphere and the clouds.

Lightning has knocked planes out of the sky, killed thousands of people, brought down electrical power grids, and caused irreparable damage to circuits, electrical systems, and critical infrastructure worldwide. Lightning is simply a beam of charged particles, just a manifestation of directed energy particles.

To Baker, charged particle beams seemed to be the ultimate weapon because they carried enormous amounts of energy vast distances: witness their destructive effects. In the early 1970s, the energy from lasers was too small to do much damage, and the energy from HPMs died off too quickly with distance to have any militarily useful effect. In those early years there were limits to laser and HPM power, and the possibility of using them as weapons was still many decades away.

But charged particle beams were different. Lightning was real, it happened all around the earth, even on cloudless days. It was one of nature's most powerful phenomena. But could lightning be harnessed? And was it possible to direct charged particle beams far enough through the atmosphere to be used as a weapon? Baker set about to make man-made lightning weapons a reality.

In a move that created anguish in scientists who had been working on nuclear effects for dozens of years, Baker turned his group's research to generating charged particle beams. The AFWL particle beam program was viewed as high-risk, high-gain. Contrary to following the popular business school guideline that "low-risk, short-term profits equal success," particle beams had such a high payoff that the air force was willing to invest scarce dollars researching this exotic field.

This high-risk, high-gain approach had proven successful for the military in the past. Investments ranged from General Schriever's ICBM program, which had allowed NASA to leverage all of the air force's development costs for rocket boosters such as Atlas and Gemini, later used for the Mercury and Gemini programs, to Admiral Rickover's nuclear reactor project, which established MIT's nuclear engineering program.

High-risk, high-gain.

Baker was playing high-stakes poker, and as science adviser to the unit charged with pulling technological rabbits out of a hat, the young Ph.D.

from Ohio was betting the future of the air force exotic weapons program on his instincts.

## Particle Beams for EMP Effects

A high-altitude nuclear explosion called Starfish conducted by the United States in 1962 had unexpectedly burned out electronics in disparate places from navy ships to kitchens. This highly energetic pulse of electromagnetic energy (called an EMP, or electromagnetic pulse) proved powerful enough to destroy electronics from hundreds of miles away, and led to a flurry of research activity trying to understand this new phenomenon.

At the time the world was rapidly developing electronic devices to control everything from communications to safety and security networks. Starfish had demonstrated that a nuclear weapon detonated high in the atmosphere might destroy electronics ranging from satellite communications to national security command and control.

As a result, much effort went into trying to understand nuclear effects, driven by both the Defense Nuclear Agency (DNA) and the British Atomic Weapons Establishment (AWE, the rough equivalent to Los Alamos or Lawrence Livermore National Laboratories). Scientists researching nuclear effects used bremsstrahlung, the radiation produced from accelerating charged particles, to simulate the X rays that might be produced from nuclear weapons.

Dr. Charles Martin, an AWE researcher, first used intensely charged particle beams to slam them into "high-Z" (high atomic number) targets such as iron to produce nuclear weapon–like levels of bremsstrahlung. Scientists in the U. S. army, navy, and air force later followed suit.

A community sprang up of nuclear-effect scientists and engineers using intensely charged particle beams to mimic nuclear effects. And, as we will see in Chapter 10, an unexpected benefit of researching new fields is that scientists discover revolutionary uses that exceed original expectations.

This foray into researching nuclear effects gave birth to the fledgling field of high-power microwaves. And this new field is arguably a much more far-reaching application than nuclear effects in this world of dwindling nuclear stockpiles.

In a perfect world, Dr. Bill Baker would have come out a hero, a risk-taking scientist who defeats all odds and whips up a Buck Rogers, blast-them-all-to-hell particle beam weapon.

In reality, he failed.

Although scientists proved it was possible to propagate electron beams in the atmosphere under the right conditions, the machines needed to generate the beams were large and unwieldy. Picture electrical equipment immersed in tens of thousands of gallons of oil, taking up a building four stories tall.

It was possible to propagate man-made lightning, but it sure wouldn't make a practical weapon. However, in conducting this research, scientists discovered something much more lethal.

The problem is that producing particle beams, much less a directed energy weapon such as lasers or HPM, is hard. Generating stable charged particle beams that won't fly apart is incredibly difficult.

Scientists had been creating low-current particle beams in accelerators for years. When the beams are confined either in a conductor (e.g., a metal tube) or by a magnetic field, the beams hold together for long periods of time. Argonne National Laboratory outside of Chicago specializes in producing specific types of particle beams and routinely accelerates them around a miles-long chamber, confining them with highly synchronized magnetic fields.

But once the restraining force is taken away, such as when the beam propagates in air, the beam flies apart and becomes unstable. The beam may lash around in the air or expand until it dissipates. In other cases, the beam may arc back and strike the machine that generated it, damaging the beam source.

Coaxing particle beams to propagate in air is like trying to harness lightning. And even if you're successful in initially directing the beam to go in the right direction, so-called beam instabilities may cause the beam to fly apart.

There are many types of instabilities, such as the fire-hose instability, where the particle beam whips back and forth like water spurting through a garden hose under high pressure.

Attempting to tailor the beam so it would be stable, scientists poured time and effort into generating complex theoretical models to determine mathematically how a beam would travel in air. The models combined Einstein's relativistic equations of motion with classical theories of electrodynamics, resulting in intertwined partial differential equations that were impossible to solve analytically.

Enormously complicated computer programs were developed to solve the mathematical models, which sometimes took many weeks to run even on the fastest supercomputer of the time. Even then, the computer models only approximated the beam, not fully taking into account the complete physical processes involved. This was limited not only because of the speed and size of the supercomputers, but because of the "scale time frame" that the computers had to model, ranging from thousandths of a billionth of a second, to many seconds—over 12 orders of magnitude, or $10^{12}$.

The simulations uncovered a curious phenomenon.

Scientists thought the particle beams would blast their way through the acceleration part of the tube. According to classical Newtonian physics, the paths of the charged particles should be constrained by the confining walls of the accelerating potential; supercharged with energy, the beam should shoot through the chamber and undergo the instabilities predicted years before.

Captain Collins Clark, a prematurely gray researcher with a quick, laughing sense of humor, proved otherwise. Clark was one of the young Ph.D.s brought in under the Vietnam draft. As a trained experimentalist, Clark could not have found a better home than AFWL. Captain Bruce Miller was another bright officer brought into the air force by the draft. Miller was a driven experimentalist who quickly awed the military brass.

In a collaboration with Sandia National Laboratory, Clark and Miller ran sophisticated experiments far beyond what their peers could even dream of. Most of their professors would never have the opportunity to participate in unique high-energy experiments, and although they were just out of graduate school, Clark and Miller were working with some of the most cutting-edge technology of the time.

After months of preparation, their experiments proved that under the right conditions, charged particle beams could propagate through air.

The air force was ecstatic. Man-made lightning appeared to be possible, and feasible particle beam weapons were just around the corner. Theoreticians such as Captain Don Sullivan from Alme's computational group were thrown on the problem in an attempt to optimize the beam and find the conditions that would make it possible to destroy an enemy warhead.

It was a heady time. In the late 1970s, with Lamberson's laser team frantically working on the Airborne Laser Laboratory, Baker's gamble to pursue particle beams seemed on the verge of making a breakthrough, a leap of energy to make a Buck Rogers death ray that would be the technological trump card in the U.S.-Soviet Cold War.

The only problem was that in some cases the particle beams started to oscillate, or wiggle back and forth. This oscillation grew and soon the beam would fly apart. Mysteriously, when this happened, electronics in the lab failed, as if disrupted by some sort of superelectronic jammer.

The experimenters tried various ways to overcome the oscillation problem. They increased the beam's voltage, then decreased it. They increased and then decreased the beam's current. They experimented with different gas densities in the chamber where the beam was created.

The theoretical team set about to simulate the beam's aberrant behavior, and their counterparts at Sandia, Los Alamos, and Berkeley joined in modeling the unusual effects. Later the researchers would learn that Soviet scientists were discovering the same curious phenomenology at about the same time.

Finally, in the summer of 1981, two years before President Reagan announced his decision to embark on an ambitious program to stop nuclear-tipped ballistic missiles in flight, another charged particle experiment failed, again due to the strange oscillations in the generation cavity. But this time, measurements set up in the lab showed a huge spike in microwave energy—high-power microwaves. They were the largest ever recorded, and they didn't come from traditional microwave tube technology that had been around since World War II. The spike was large enough to destroy several electronic instruments in the lab, and when a detailed analysis was made, it was traced back to the particle beam oscillations.

The researchers discovered that the particle beams oscillated at microwave frequencies. Although the technical details are obtuse, in small, localized patches, the charged particle beam moved back and forth so fast that it created microwaves hundreds of times greater in power than what could be created by traditional tubes such as those found in radar, aircraft, or microwave ovens.

These microwaves were an unintended by-product of the charged particle beam experiments. When conditions were optimized, a virtual cathode oscillator, or vircator, was discovered that created microwaves hundreds and even thousands of times more powerful than the state-of-the-art tube technology that had evolved from the radars used in World War II.

Thus in an unexpected discovery, high-power microwaves were made possible; not through the traditional evolutionary means of technical improvement, but rather from a totally unexpected direction, coming out of left field.

Microwave tube technologists had nothing to do with this invention. In an unexpected twist, high-energy physicists were key in creating the weapon of the future, a discovery that no one could have foreseen. All because of a radical change in research championed by Bill Baker.

Who would have thought?

The first documented use of directed energy. In 212 B.C., Hippocrates, commander of the Greek forces, focuses sunlight with mirrors, setting fire to sails of the Roman fleet at the siege of Syracuse. This work, from the Stanzino delle Matematiche in the Galleria degli Uffizi (Florence, Italy), was painted by Giulio Parigi (1571–1635) in 1599–1600.

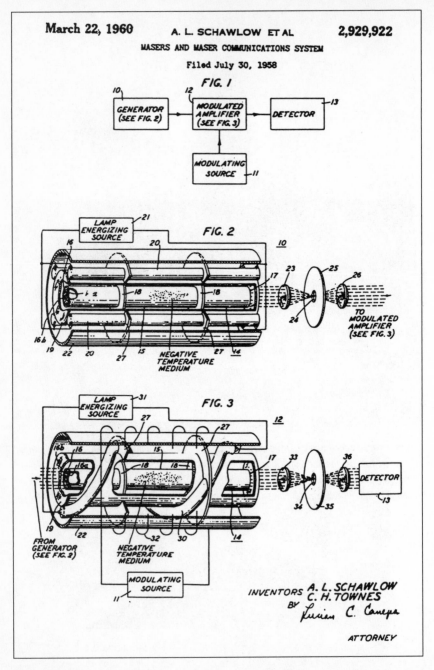

Diagram from the first patent for a microwave laser (maser) by Arthur Schawlow and Charles Townes.

Dr. Diana Loree, the present Active Denial system (ADS) program manager, and Dr. Kirk Hackett, the first ADS PM, in front of the Humvee-mounted ADS advanced concept development technology (ACDT) unit built by Raytheon for the air force. Photo courtesy Air Force Research Laboratory

Artist's conception of an airborne Active Denial system aboard a C-130 gunship. An airborne ADS could provide a means to employ nonlethal weapons at distances much farther than a ground-based ADS. Enhanced photo courtesy Air Force Research Laboratory

The Active Denial system 0 source used at the test site on Kirtland AFB, New Mexico. This unit was used to conduct several hundred exposures of Active Denial at distances greater than small arms fire, proving that it was possible to assess intent and control crowds at a distance using nonlethal force. Photo courtesy Dr. Kirk Hackett

Active Denial test subject 1, Lieutenant Colonel Chuck Beason, before being subjected to the full-body ADS effect at Kirtland AFB, New Mexico. Photo courtesy Dr. Kirk Hackett

$CO_2$ gas dynamic laser (GDL): Experimental laser device (XLD) 1, Pratt & Whitney test site, Florida Everglades. Photo courtesy Pratt & Whitney

The navy–ARPA chemical laser (NACL) was the first to lase with more than 100 kilowatts of power, achieving its first light in 1975. Photo courtesy John Albertine

Mid-infrared advanced chemical laser (MIRACL), at White Sands Missile Range, is a combustion-driven deuterium fluoride (DF) chemical laser that first lased in 1980. A megawatt-class laser built for the SeaLite program, it was the Defense Department's largest laser until the ABL fired. Photo courtesy John Albertine

The Alpha laser, built for SDI space-based laser (SBL) program, was a combustion-driven, megawatt-class hydrogen fluoride (HF) chemical laser that achieved "first light" in 1987. Note the cylindrical gain generator, making this design efficient for being packaged for spaceflight. HF laser radiation won't propagate in the atmosphere because of absorption by water molecules, so it was thought to be ideal for striking ballistic missiles that rose into space. Photo courtesy TRW (now Northrop Grumman)

San Juan Capistrano laser test facility, home of the NACL and Alpha lasers, among others. Photo courtesy TRW (now Northrop Grumman)

SeaLight beam director (SBD). The navy's effort to demonstrate a large ground-based laser culminated with the MIRACL (mid-infrared advanced chemical laser). Located at the White Sands Missile Range high-energy laser system test facility (HELSTF), MIRACL is a megawatt-class combustion-driven DF defense SeaLight program. It was the Defense Department's most powerful laser until the ABL fired. Photo courtesy John Albertine

Army mobile tactical unit (MTU), a 50-kilowatt $CO_2$ electric discharge laser (EDL) mounted on a Marine M113 amphibious troop carrier. The MTU shot down a drone aircraft at short range in 1976. Photo courtesy U.S. Army

The Air Force Airborne Laser Lab (ALL) was a USAF KC-135 (Boeing 707) $CO_2$ gas dynamic laser (GDL) operating at a classified power level. It shot down drones and air-to-air missiles from the late 1970s to the early 1980s. The aircraft is now at the Air Force Museum in Dayton, Ohio. Photo courtesy Air Force Research Laboratory

The joint U.S. Army and Israeli tactical high-energy laser (THEL) system was built to provide defense against short-range artillery rockets. A submegawatt class DF laser, over 30 Katysushi rockets and artillery rounds have been shot down at White Sands Missile Range. A mobile version, MTHEL, is under construction. Photo courtesy Northrop Grumman

Pointer-tracker from army's THEL. The pointer-tracker follows the target and accurately points the laser. Photo courtesy Northrop Grumman

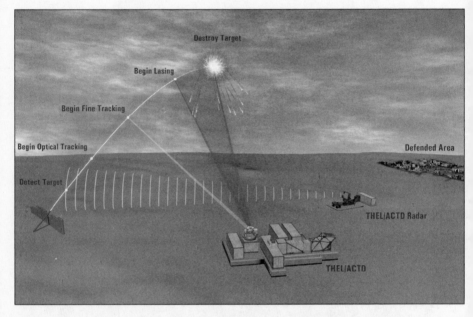

Artist's conception of the THEL being used in combat. A radar tracks the incoming target and relays the targeting information to the THEL pointer-tracker. The THEL irradiates the target with laser radiation until the target's warhead explodes. Picture courtesy Northrop Grumman

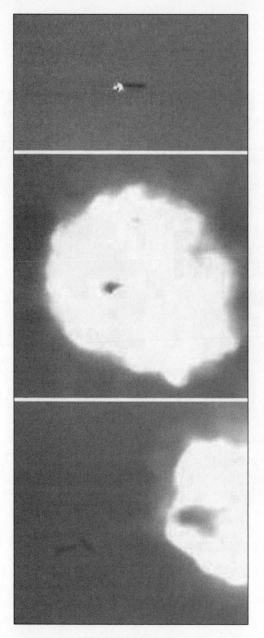

A Katyusha rocket (upper panel) is lased by the army's THEL (tactical high energy laser) at White Sands Missile Range. The DF laser heats the rocket, causing the high-explosive warhead to detonate. Photo courtesy Northrop Grumman

A pressurized rocket booster exploding during a test at White Sands Missile Range when hit with a high-power laser. The laser softens the missile skin, allowing the internal pressure of the missile fuel tank to rupture and explode. This is the "kill mechanism" the ABL will use. Photo courtesy Missile Defense Agency

The first flight of the USAF airborne laser after being modified at Boeing's Wichita, Kansas, facility. Photo courtesy USAF ABL System Program Office

The airborne laser in test flight in 2004. Photo courtesy USAF ABL System Program Office

The airborne laser in test flight in 2004. Photo courtesy USAF ABL System Program Office

# 9

# The World's First Force Field:
# The Active Denial System

---

MAY 7, 2001, was a typically spectacular Albuquerque day. The sky was cloudless, deep blue; visibility was over 100 miles, revealing Magdalena Ridge to the south and Mount Taylor to the west. It was a calm day, without the spring winds that whip across the high desert that time of year—and did not reflect the policy winds storming a thousand miles away to the east.

It was the perfect day in New Mexico to perform the Active Denial experiment, the world's first attempt to project what the media would later call a nonlethal force field at a distance greater than small arms fire.

Three weeks before, on April 20, in Washington, D.C., a Department of Defense official approved Active Denial by signing a document entitled "Unclassified Human Use Protocol for Large Spot Human Pain Intolerability Experiments." The air force surgeon general had approved the project on January 22. But since the experiment involved a new technology—one that was potentially a hotbed of media criticism—the final approval had to come from the undersecretary of defense, who was responsible for all military acquisition, logistics and technology.

After years of classified research, the final okay had been given to allow the military to project directed energy in the form of millimeter waves in the field (i.e., outdoors) in a realistic, nonlaboratory environment against humans.

If successful, the test would prove that millimeter waves might someday be used as a nonlethal weapon. But the path to this test involved over a decade of research and dogged determination to convince the authorities that Active Denial was a viable nonlethal directed energy weapon, not some sort of crazy death ray that had sinister, long-term effects.

Unlike lower-frequency electromagnetic radiation such as radio waves and microwaves, millimeter waves do not penetrate far into the body. Active Denial technology (ADT) uses millimeter waves centered around 95 gigahertz (written as GHz, where a GHz frequency is 1,000 megahertz), corresponding to an electromagnetic wave a little over 3 millimeters in length, or about a tenth of an inch.

This wavelength is 40 times smaller than the microwaves generated by home microwave ovens and is absorbed into the first one-third of a millimeter in the skin.

Thus millimeter waves do not directly interact with nerves or pain receptors, nor do they affect internal organs such as the brain. Instead, millimeter waves are converted into heat, which affects the pain receptors and is experienced as intolerable pain. People exposed to millimeter waves experience a "flee" response and instinctively try to get away from the waves. When the millimeter waves are turned off, the pain immediately subsides and there is no physical damage.[1]

Unlike potentially harmful charged particles or high-frequency electromagnetic radiation such as X rays, millimeter waves are nonionizing; they do not cause chemical changes when they are absorbed. Rather, millimeter waves transfer their energy in the form of thermal energy, or heat.

Photons of millimeter waves are less energetic than photons of visible light by a factor of 10,000, meaning that there is a much lower probability of inducing any chemical change than through visible light. Millimeter wave studies coordinated through the Tri-Service Electromagnetic Radiation Panel (chartered through the deputy undersecretary of defense for environmental security) over the past twenty years have published results in peer-reviewed journals showing that millimeter waves are noncarcinogenic.[2]

Critics maintain that a longer-term effect ("unknown unknowns") may not show up for decades, pointing out that carcinogenic effects in servicemen exposed to nuclear tests did not appear until much later. However, medical researchers at Brooks City Base in San Antonio, Texas, maintain that the nonionizing nature of millimeter waves will preclude this from happening. Their claims are backed up with over a decade of research that has been openly published in the literature.[3]

In the past, researchers had only exposed small patches of skin to millimeter waves, and then under tightly controlled laboratory conditions. The "repel" or "flee" response, now known as Active Denial, had been documented over the past decade at the Air Force Research Laboratory's Directed Energy Bioeffects Division, but only on skin areas that ranged in size from the tip of a pen to a few centimeters in diameter. In these tests, each subject's increase in skin temperature, heart rate, respiration rate, and other diagnostics were closely monitored to quantify both the degree and the time it took to achieve the flee effect.[4]

The author was allowed to experience the entire array of millimeter sources, including one of the first tests conducted in the field. Someone who is subjected to millimeter waves feels intense heat, as if from a supercharged oven. For the smaller-diameter millimeter wave beams (on the order of the tip of a pen), the sensation is not unlike a hot pinprick; the larger beams produce a faster and vastly more painful sensation.

Armed with a wide background of medical research and small skin patch test data, air force researchers felt confident that it would be safe to expose a person's body to millimeter waves. But many questions remained: Would the subjects feel a sensory overload? Would their physiological system slam into high gear trying to flee the exposure? Or would the millimeter wave energy somehow feel as if it had been spread out over the body? And would the exposure not be perceived as pain? Or would the subjects go into convulsions and not be able to function?

The Active Denial program had to answer these questions before it could continue. The years of small-scale experiments on humans, the decades-long research program on animals, and other laboratory experiments were not sufficient to answer these crucial questions. Active Denial was at a juncture; to continue, a human use protocol had to be approved.

In the fall of 2000, Active Denial technology (ADT) was classified as a Department of Defense secret program, and researchers needed official permission to conduct classified experiments on humans. Under presidential directive, that permission was only authorized to be granted by an agency head and could not be delegated. In this case, the agency head was the secretary of defense, William Cohen.[5]

Department of Defense regulations dictated that all classified experiments had to be reviewed by an institutional review board located at Wright-Patterson AFB prior to receiving the secretary's approval. That process must be conducted in a 30-day window designed to provide the secretary with the most up-to-date advice on the veracity of experiments shrouded under the veil of secrecy. Twice the Active Denial researchers tried and failed to meet that 30-day window because of the time it took to conduct the myriad reviews through the bureaucratic layers in the Pentagon.[6]

After years of conducting research, studying the effects of Active Denial on animals, detailing the small-scale effects on humans, publishing their results, and building what seemed to be a solid scientific case for the first directed energy nonlethal weapon, the program ground to a halt. There seemed to be no overcoming the bureaucratic stalemate until the program was briefed to the same man who helped nurture a similarly promising DE program—ALL, the Airborne Laser Laboratory—years before when he was secretary of the air force: Hans Mark.

In 2000, as assistant to the secretary of defense, Dr. Hans Mark held the position of DDR&E, director of defense research and engineering. Mark was not only Pentagon spokesman for science and engineering, but he oversaw the Defense Department's $40 billion budget in research, development, testing, and engineering.

Before coming to DDR&E, Mark held numerous influential and distinguished positions: director of NASA Ames Research Center, undersecretary of the air force while simultaneously serving as director of the National Reconnaissance Office, and secretary of the air force before becoming deputy administrator of NASA. From there, he went on to become chancellor of the University of Texas system.[7]

As air force secretary in the early 1980s, Mark supported the Airborne Laser Lab, an experimental $CO_2$ laser in a reconfigured KC-135, a military version of a Boeing 707 that eventually shot down air-to-air missiles fired at the plane. In the early 1970s Mark was influential is establishing that program.[8]

Mark did not shy away from high-risk, large-scale technical programs. He was demanding, and program officers who were used to overwhelming their nontechnical audiences with an onslaught of information found their match in Mark's detailed questioning.

Mark was well aware of Active Denial. Early in 2000, Mark criticized the technical conclusions of the Active Denial technology (ADT) program manager, Dr. Kirk Hackett, during a spirited Pentagon briefing. This did not sit well with Hackett, an MIT-trained Ph.D. physicist, who had established the ADT program years before. Mark refused to accept either the range or the effect of the Active Denial system (ADS) that the air force proposed to build, and bluntly declared that the system would never work as advertised. Not intimidated by Mark's high-level position, Hackett stood his ground and heated words were exchanged as each man challenged the other's technical logic.[9]

Soon afterward Hackett invited Mark to the Human Effectiveness Lab at Brooks AFB. A cadre of DE researchers made the trip, accompanied by Dr. Bob Peterkin and Dr. Tom Hussey, at that time senior scientist of the Air Force High-Power Microwave Division. Also present was Dr. Mike Murphy, Jim Merritt, Dennis Blick, and Lieutenant Colonel Chuck Beason from Brooks AFB, as well as Major General Dick Paul, commander of AFRL, Dr. Earl Good, director of AFRL's DE directorate, and Dr. Bill Baker. The team led Mark through a detailed analysis to convince him of the scientific basis behind ADT.

The Brooks researchers sat Mark down in a chair and exposed the distinguished scientist to a small-diameter millimeter wave beam to allow him to personally experience the beam, simulating the ADT effect at a distance.

When the switch was thrown, Mark shot out of the chair like a rocket. In the words of one researcher, "I never saw a political appointee move so fast in my life!"[10]

Mark became a believer, convinced the Active Denial effect was real. He apologized to Hackett and became one of his biggest supporters. Soon Mark flew to Albuquerque to confer with the senior leadership of the air force's Directed Energy Directorate.

That night, Mark, Dr. Good, the author, and several other researchers had dinner by the Albuquerque airport at a quiet seafood restaurant called the Yacht Club, an inside joke to New Mexicans, as the entire state had hardly enough water to launch a yacht, much less support a yacht club.

After receiving a program update from Hackett over dinner, Mark pounded on the table, startling several diners. He was livid. Now that he appreciated the nonlethal weapons capability of ADT, he was infuriated by the bureaucratic brick wall Hackett faced.

The Pentagon bureaucracy seemed to be doing everything it could to thwart human testing. Exposing people to ADT was critical to determining if this technology was going to be a savior for the military or if it should be shelved, as other nonlethal techniques had been in the past. The brick wall was the stringent timetable erected by the Clinton administration intended to ensure that classified tests were not misused—a good goal. But unfortunately the unbending policy impeded promising technologies, rather than helping them.[11]

Over dessert, Mark laid out a plan to get around the unrealistic constraints. He reasoned that if ADT were unclassified, human testing would be open to public scrutiny. Then the experimentation phase would be quickly approved through the Pentagon.

That night, Mark directed Dr. Earl Good, director of the Air Force Research Lab's DE Directorate, to declassify ADT: "This technology must be deployed. It's much too important for our nation."[12] Subsequently Mark called the Phillips Research Site commander, also the senior AFRL military officer at Kirtland, several times over the ensuing month to see if ADT had been declassified.

Finally, on December 8, 2000, the fact that a directed energy nonlethal weapon research program existed became unclassified. As expected, approval for the human testing protocol was quickly approved up the chain of command, without fanfare.

The Pentagon worried that adverse publicity would short-circuit Dr. Mark's plan to conduct experiments on humans to determine the military utility of ADT. A huge public outcry was anticipated, and it was feared ADT would be overcome by negative publicity.

Behind the scenes, the DE public affairs official, Rich Garcia, himself a retired colonel, coordinated a detailed briefing with the Marine Corps, which ran the Active Denial system program out of its joint nonlethal weapons office. Colonel George Fenton, a wiry, no-nonsense Marine, directed the joint office. Fenton worked with the Pentagon to lay out a measured, media-friendly plan to introduce the public to ADS (ADT refers to Active Denial technology; ADS is the Active Denial system, the actual nonlethal program).

But on February 23, 2001, after a reporter was mistakenly allowed to remain at a "for official use only" briefing on a war gaming session, the *Marine Corps Times* broke the story to the public—exposing ADS to a nation that just 60 years earlier had panicked when it heard a radio broadcast about death rays from Martians.[13]

The air force and Marines issued a joint press release, but the news spread rapidly.[14] That night, millions of people listed to Jay Leno as he joked about the new nonlethal weapon, and how U.S. soldiers would have to procure rotisseries to check if the enemy was "done."

The first full-exposure human ADT test took place on a remote corner of Kirtland Air Force Base, five miles south of the Albuquerque city limits. In addition to the armed air force security police manning the entrance to Kirtland AFB, the test site was guarded by a series of spotters, as well as two guards at additional checkpoints. Dark red badges emblazoned with the head of a fire-breathing dragon and the words FEEL THE HEAT, were issued to those authorized to view the test.[15]

Test subject number one was selected from a pool of over 80 volunteers. The test location was located east of the Active Denial source, at a distance outside of the range of small arms fire, over 700 meters away.[16]

The Active Denial source, known as System 0 (shown in Figure 9.1), was housed in an air-conditioned portable trailer that sat on a small rise and looked more like a rundown storage shack than the world's most sophisticated nonlethal weapon.

Next door, another rundown trailer held the tracking equipment and operator console. Visible and infrared images displayed from a telescopic lens enabled the operator to view the test.

Undressing in an RV trailer, test subject one—Lieutenant Colonel Chuck Beason, Dr. Kirk Hackett's longtime cohort championing Active Denial—stripped down to his underwear, pulled on a ratty bathrobe, and stepped into the glaring New Mexico heat. Then he was led across the desert floor to the test area. As a joke, Hackett had bought Beason a leopard-skin thong to wear for the test, but, after pondering what the official pictorial record would reflect of test subject one, the lieutenant colonel refused.[17]

The test area consisted of a canvas army tent with an open flap. On either side of the flap, insulated barriers opaque to millimeter radiation were erected so that doctors, psychologists, and other emergency personnel could stand close to the subject and observe the test.

The initial testing protocol only allowed back exposures. Even though clothing is transparent to millimeter waves, the experimentalists needed to measure the rise in skin temperature using infrared radiation, which would be trapped by clothing.

Figure 9.1    Active Denial System 0, courtesy Air Force Research Laboratory. Note the large directional antenna on top of the trailer. (Photo courtesy of Dr. Kirk Hackett)

Test subject one turned his back to the Active Denial source and shrugged off his bathrobe. A foam-padded mat was spread out around him, in case he fell to the ground.

The test protocol dictated that the beam would remain on for a short time (the actual number is sensitive) and that the subject would try to remain still as long as possible. If the pain from the beam became unbearable, the subject was instructed to either fall forward onto the mat or leap to the side, taking him out of the beam's path.

After baseline measurements of the subject's vital statistics were taken, the countdown began. When the count reached zero, test subject one instantly felt as if a huge oven door had been opened behind him.

He felt the heat soak into his body, covering his head, his neck, his back, his arms, his buttocks, the back of his thighs, and all the way down to his heels. Every spot on the backside of his body was immersed in a sensation of sudden, incredible hotness.

But unlike an oven where the heat seemed to reach a plateau and level out without getting any hotter, the sizzling feeling grew more and more intense.

Nearly instantly, the subject yelped and sprang to the side, into the shadow of the protective shield.

At the ADS source, the operator saw the subject jump out of view of the beam and immediately released the trigger. The beam had been programmed to shut down within a specified time to prevent any unintended exposure, but the subject had not stayed in the test area long enough to complete the first full human exposure.

Cheers erupted from the tent area. The subject was ecstatic along with the rest of the experimental crew, elated that the experiment worked.

Doctors rushed to the test subject and readministered the tests they'd conducted just prior to the exposure. Except for an increase in heart rate from the excitement of the test, the subject showed no signs of stress. The subject's skin was slightly red where the beam had struck, but the redness went away after a quarter hour; later, researchers determined the redness was due to blood rushing to the affected area. The doctors kept the test subject under observation to ensure that there were no unintended symptoms.

In subsequent testing, many VIPS, including several four-star generals, political appointees, and a former air force chief of staff, participated as test subjects.[18] As this book goes to press, over 1,000 subjects have been exposed to the Active Denial system.

A measured test protocol allowed frontal exposure to the millimeter waves only after the back exposures were thoroughly reviewed. The database of exposure shows no adverse effects, consistent with the data generated by decades of small area exposure.

The Active Denial testing grew more sophisticated, with exposures made in realistic scenarios and environments to assess the military utility of using ADT. For example, one scenario consisted of an aggressive crowd converging on several placid people. The ADS operator was able to expose each individual aggressor, called the red force, and push back the crowd without exposing any innocent bystanders (also test subjects) to the ADT effect.

Other scenarios included aggressor personnel intermingled with innocent bystanders, called the "blue force"; another had aggressors stealthily approaching a landmark. In most scenarios, the ADS operators successfully discriminated between the aggressors and the bystanders, and drove the aggressors away. When blue forces were hit, the operators changed their tactics and made ADS that much more effective.

While the Active Denial System 0 was being used to conduct the human protocol testing, Raytheon, the integrating contractor for the air force's Humvee-mounted ADS 1, was completing the installation of the gyrotron, antenna, power supply, and remaining infrastructure onto the vehicle. Raytheon vice president for directed energy weapons Mike Booen (a retired air force colonel who ran the air force's airborne laser program from 1997 to 2001) oversaw delivery of ADS 1 to the air force in September 2004.

After ADS 0 successfully demonstrated the Active Denial effect at distances greater than 700 meters, Dr. Kirk Hackett, the program manager and acknowledged father of ADS, was transferred to Washington, D.C., where he advised the director of defense R&D on weapons science and technology. Dr. Diana Loree, his deputy and a Ph.D. in electrical engineering from Texas Tech University, stepped up to take his place with Active Denial System 1.

Dr. Loree oversaw the ADS program through assaults on its budget as it navigated through the scrutiny of DoD's advanced technology concepts demonstrator (ACTD) process. ATCDs are partially funded out of the Department of Defense programs from the office of the undersecretary of acquisition, technology, and logistics (AT&L).

Being designated an ACTD was good for the program because ACTDs have high-level support and the undersecretary's attention. The undersecretary for AT&L is nominally the third most powerful person in the Department of Defense (depending on the administration, this position has vacillated in priority between number 3 and 5). And if an

ACTD is high on the undersecretary's priority list, by definition it's high on everyone's list.

A "lead service," or primary military service contact, is responsible for shepherding the ACTD through the demonstration process and must find additional funds to support the ACTD.

For ADS, the air force was chosen as the lead service, and the joint nonlethal weapons program office was tapped to provide additional funding. So at first blush, ADS being designated as an ACTD seemed like a good thing.

However, being an ACTD can also be bad.

The bad part is that if the ACTD runs over its budget, the lead service (in this case, the USAF, which was also saddled with supporting another major directed energy program, the airborne laser) was responsible for making up the funding shortfall. And because the USAF had not previously fully supported the ADS mission, there was no war fighter support of this transformational technology.[19]

While representing the USAF in supporting ADS as the lead service, Dr. Loree did not have the full backing of senior USAF leadership.[20] With the exception of Dr. Hans Mark, no senior person in the Department of Defense was willing to step up and say that this might be the best way to save American lives without killing the enemy. For years, even the air force science advisory board pushed to accelerate Active Denial to combat, but its pleas went unheard.[21]

DoD's joint nonlethal weapons directorate has invested about $40 million toward Active Denial technology. At the time, the air force's interest was based on ADS being a ground weapon, thus having limited appeal to the world's premier air power.

During these pivotal times for ADS—just months after the devastating 9/11 attacks, and the commitment from Secretary Donald Rumsfeld to "transform our forces"—the majority of air force leadership did not support this revolutionary technology.

The USAF funds many projects and has myriad priorities, but the agency also has a yearly operating budget approaching $100 billion. And to not find $50 million over five years—a hundredth of a percent of that budget—to support what has been recognized as "indispensable for saving lives," is questionable at best. Especially when the next generation of that weapon might be used in the air and might well prove revolutionary for controlling large enemy crowds at a distance.

Leading any acquisition program is tough, regardless of the level of support. But working without the necessary funding or encouragement by senior leaders, in contrast to what was displayed for General Lamberson's Airborne Laser Laboratory in the late 1970s, is a tribute to Dr. Loree's stamina and persistence.[22]

It also shows the difference between the leaders of yesterday and today. For example, General Lamberson's ALL program was supported by myriad generals and political appointees, including the secretary of the air force. They realized that the ALL would demonstrate the feasibility of someday employing an airborne laser weapon. And they realized that the investment would take years to come to fruition. Contrast that support to the present uphill battle for ADS.[23]

To be fair, the lack of support for the Active Denial program did not occur on the watch of the current senior leadership. But it should behoove our present leaders to learn from past giants such as General Arnold and Dr. Von Karman, the first chairman of the air force scientific advisory board, when they realized today's investments reap tomorrow's dividends.[24]

So, after a decade of research and bureaucratic battles for its survival and despite the lack of high-level support, the world's first deployable antipersonnel directed energy nonlethal weapon is ready for its ultimate test—to see if it can really save lives at the speed of light.

# The Birth of the Airborne Laser

It isn't very often an innovation comes along
that revolutionizes our operational concepts,
tactics and strategies. You can probably name
them on one hand—the atomic bomb, the satellite,
the jet engine, stealth, and the microchip.
It's possible the airborne laser is in this league.

— AIR FORCE SECRETARY SHEILA WIDNALL,
*Airman*, April 1997

ALL MAJOR WEAPONS SYSTEMS have a champion and a begin-
ning, and the airborne laser (ABL) had both, as well as a legacy in the air
force Airborne Laser Laboratory (ALL). The ALL story, documented in a
book by Robert Duffner, chronicles the difficulty in fielding a high-
power laser weapons system on a flying platform.[1]

Several major differences exist between the ABL and ADS (Active Denial system, the nonlethal millimeter wave antipersonnel DE weapon
examined in Chapter 9). First, ABL had its roots in the clearly traceable
timeline of moving from a ruby laser to a gas dynamic laser, in pursuit

of the Buck Rogers/Star Trek expectation of blowing an enemy to hell, not in the esoteric, intimidating world of nuclear effects and high-energy particle accelerators like HPM.

Plus, the ALL captured the imagination of the USAF senior leadership. And this was the major discriminating difference between the success of the USAF ALL and subsequent ABL, to the struggle of the present-day high-power microwave program and the other military high-energy laser programs.[2]

In a seminal meeting in 1968, the air force secretary, chief of staff General Jack Ryan, the commander of USAF Systems Command, and other less illuminary generals attended a USAF science advisory board meeting where Edward Teller (the father of the H-bomb) gave an impassioned plea for an airborne laser weapon.[3] This gathering of high-level officials purely for a science and technology program had never occurred before. Those present were struck by the support and vision of all leaders. Contrast this with the seeming indifference of present-day senior leaders for revolutionary technology.[4]

Although based on $CO_2$ laser technology, the ALL demonstrated that a weapon-level laser system could be deployed on board a military aircraft and that the associated problems of jitter, instability, air turbulence, high-frequency vibrations, pointing and tracking, exhaust, and thermal expansion of the beam could be compensated for while shooting down targets—including air-to-air missiles launched at the ALL itself.[5]

Although the ALL was a successful test, it was just that—a test, a demonstration. It was not a weapons system, nor was it a modern version of the Defense Department's ACTD (advanced concept technology demonstrator—a program designed to leave a "residual" war-fighting capability when it was complete, such as the wildly successful JSTARS (joint surveillance and target attack radar system), which was pushed into service during the first Gulf War to track enemy combatants on the ground).

The ALL was a demonstration program, fielded to prove the necessary but difficult tasks listed above. In addition, the ALL $CO_2$ laser did not have enough power or range to qualify as an operational weapons system.

But to be fair, demonstrators don't have to do that, and they shouldn't. Otherwise a phenomenon known as requirements creep will slowly envelope a program and suffocate it.

A good example of requirements creep happened to the national space plane (NASP), President Reagan's attempt to build an orbital Orient Express. The vision in the late-1980s was to build a spacecraft capable of taking off like a plane, accelerating to orbital velocities, and landing on runways on the other side of the globe within an hour and a half. The NASP was feasible according to the laws of physics, but a nightmare for engineers to make possible, similar to Dr. Bob Forward's example of what's possible versus what's probable described in Chapter 7.

Instead of requiring the NASP to merely achieve orbit and function as a plane—the minimum requirements any sane designer would make on a first-generation demonstrator—political agendas began to pile up, and one by one additional requirements were added to the NASP: it had to carry a minimum weight to orbit so that the military could use it; it had to carry a minimum volume to orbit so that NASA could use it; it could not go supersonic before reaching a certain altitude so that the environmentalists/government watchdogs would not be offended; parts of the NASP had to be constructed in every state of the union so that the entire Congress would support it; the list went on.

As you can see, with each additional requirement, the probability of failure grew. And eventually the NASP was canceled. The point is that constructing the airborne laser lab as a demonstrator made good sense. It proved that a laser weapons system could be installed on a flying platform.

But so what? Why should the United States spend billions of dollars on a laser weapon that can shoot down air-to-air missiles? ALL did not perform a unique mission. F-15 and F-16 fighters can engage enemy fighters much more agilely, and they cost a heck of a lot less than ALL.

In the early 1980s, when ALL conducted its first tests, F-16s were just being introduced, with the first F-16 delivered to the 388th Tactical Fighter Wing located at Hill Air Force Base, Utah, in 1979. F-16s were being built by the hundreds, and their unit cost was on the order of $15 million in constant FY98 dollars. Armed with their own air-to-air missiles, a squadron of F-16s could engage an enemy far beyond the distance that ALL could lethally project its laser power.

So why use an expensive laser to do what an inexpensive fighter can do? In a way, the ALL was a solution looking for a problem—a unique problem that could only be solved with a laser.

The problem was to find a weapons-class laser, married with a compelling problem that no other weapons system could solve.

And then came the first Gulf War.

Scud missiles, which are based on V-2 rockets, are not sophisticated weapons. They don't have much range (60–800 kilometers), and they're not very accurate.[6] In fact, they may or may not travel in the general direction they're intended to go.

But Scuds can carry high explosives, chemical, biological, or even nuclear warheads. And they can strike terror into the hearts of nations, as demonstrated by Saddam Hussein's ruthless pummeling of U.S. troops, Saudia Arabia, Kuwait, and Israel during the first Gulf War.

Worse yet, there was no defense against the Scud launches.

None.

Zero, and that was from the nation that had the greatest air force on the planet.

This was partly because a former USAF chief of staff purged electronic warfare (EW) capability from the air force arsenal, which would have allowed detection and mitigation of that threat, and partly because of the covert and stealthy nature of Iraq's Scud forces.

In any event, Scuds were launched against U.S. and allied forces in the first Gulf War. And even with billions of dollars of overhead assets and engaging the most sophisticated air force in the world, once one of these tactical ground-to-ground missiles was launched, even the touted Patriot antimissile system (which after the war was discovered to have downed *no* Scuds) could not stop this ballistic missile threat.

The United States, a nation that outspends the next 15 countries combined in military defense, had nothing that could stop this relatively low-tech threat. It was a threat that had been initiated in World War II Germany and perfected in 1990 Iraq.

It stretched the capability of the U.S. Defense Department just to locate a launch, and hundreds of F-15 sorties (missions) never detected a single Scud launcher erecting to launch.[7]

So what was the nation to do? Stopping Scuds seemed an impossible task. The Scud cycle time was minutes—just under the time it took for the United States to acquire a launch site and scramble fighters to destroy the launchers.

Fighter jets can only fly hundreds of miles an hour. Even the most sophisticated jet in the world, the new F-22 Raptor, which cruises faster than the speed of sound, takes tens of minutes to travel the hundreds of miles needed to destroy an adversary. The F-22 might use a long-range supersonic air-to-ground missile to destroy the Scud launcher, but that missile still takes several minutes to arrive. And an F-22 costs a good fraction of a billion dollars, so this is no bargain-basement F-16.

Now, what would happen if these Scud-like missiles were suddenly deployed by the bad guys near Europe, next to our allies? Or even worse, on a disguised freighter a few dozen miles off the coast of New York or Los Angeles, where we couldn't afford the luxury of waiting even a few minutes to destroy them?

How fast is "fast" when it comes to deploying an F-22 or even a squadron of them, when we're not even sure where the threat is? And what would be the cost of deployment?

The United States faced a decision. It could build thousands of F-22-like fighters and station them every hundred or so miles apart in enemy (or our own) territory to meet this new threat—or it could resort to an entirely different paradigm that deployed its lethal weapon over a great distance nearly instantaneously and do it at the speed of light.

After the first Gulf War, USAF chief of staff General Ron Fogleman and secretary of the air force Sheila Widnall were in violent agreement. The Scud problem could be solved with a laser weapon, mounted on a platform that engaged the ballistic missile as it broke above the cloud layer, typically 40,000 feet above sea level. There the missile would still be accelerating, burning its fuel. The missile would be bright to every tracking system known, and since it would still be accelerating, it would be vulnerable to perturbations; a missile is delicate and susceptible to tiny variations in its flight conditions. Any small change in its trajectory, aerodynamics, or thrust would destroy its ability to reach its target—which is a good thing.

General Fogleman and Secretary Widnall, a professor at MIT before being tapped as air force secretary, were reminded of the possibility of fielding a high-tech solution to the Scud problem by the air force R&D arm situated in Dayton, Ohio, at the Air Force Material Command headquarters.

The ABL solution grew out of a multipronged lobbying effort by industry and government protagonists. Fogleman and Widnall quickly saw the benefit of fielding a strategic laser system. Although there was much skepticism in the "traditional air force," Fogleman, as the senior air force four-star general, moved fast when he spotted a trend.[8]

## The Air Force Stands Up the ABL SPO

In 1987, Colonel Dick Tebay, a 1967 USAF Academy graduate and B-52G pilot, was assigned to Wright-Patterson AFB in Dayton, Ohio, to work in Aeronautical System Center's simulator program office. The Simulator Systems Program Office (SPO) oversaw acquisition of all USAF simulators used at flying bases throughout the world. The simulators gave pilots the experience of flying a plane.

Tebay had spent his career flying; that was why he had joined the air force, not to buy things. Although the simulator office was a nonflying job, it was about as close to flying as he could get. The assignment was called a career-broadening opportunity, and Tebay looked forward to one day heading back to the cockpit and ending his AF career in a flying job.

Which was why Tebay was more surprised than anyone when in the late fall of 1992, he received a call from Lieutenant General Barry, commander of the Space and Missile Center located in Los Angeles. Space and Missile Center was the sister acquisition center to Wright-Patterson's Aeronautical Systems Center, but instead of buying airplanes, SMC bought space hardware such as rockets, satellites, and space communication equipment.

General Barry's conversation was curt: "Dick. Are you interested in starting up the airborne laser?"

"I don't know anything about lasers."

"Don't have to. There'll be a lot of smart Ph.D.s who can tell you about that. I need a colonel to head up a laser weapons SPO. Are you interested?"

"Is there any money in the budget?"

"No. We're putting it under a weapons technology PE [program element].[9] It's not an official program yet, but it will be soon. Let me know your answer in a week." And he hung up.[10]

Thirty days later, in December 1993, Colonel Tebay arrived at Kirtland AFB, Albuquerque, home of the USAF Phillips Laboratory. Formerly known as the Air Force Weapons Laboratory, the Phillips Lab was the birthplace of COIL (chemical oxygen-iodine laser) and adaptive optics, and the future home of the airborne laser. Tebay's family followed him from Dayton to Albuquerque later in 1993.

In the meantime, Tebay had a colossal chore. General Fogleman and Secretary Widnall had rammed the ABL through the air force system and expected the ABL SPO to hit the ground running. They assigned Dr. Steve Lamberson, a retired lieutenant colonel and former head of the Phillips Lab's COIL program, to be the ABL chief scientist—one of the "smart Ph.D.s" that General Barry had promised Tebay. Steve was a laser weapons community insider. His father was Major General (Dr.) Don Lamberson, who had been responsible for fielding the USAF Airborne Laser Laboratory that had successfully shot down air-to-air missiles over a decade before.

Working with the Phillips Lab commander, Colonel Rich Davis— a young laser physicist who would eventually go on to become a two-star major general—Tebay gathered together the Phillips researchers and their contractors who had been working on the fledgling laser technology since the 1960s. Without spending a dime, he arguably had the best talent in the world to build a laser weapon demonstrator.

That was when he discovered that his biggest problem was not technical but cultural.

Tebay had been charged by General Barry to build a laser weapons system, not a demonstrator. Laser weapon demonstrations had already been made. Project Delta proved that lasers could shoot down unmanned drones in the 1970s; the navy's ground-based UNFTP (Unified Navy Field Test Program) shot down TOW missiles in 1978; the Airborne Laser Lab proved that it was possible to track and shoot down air-to-air missiles from an airborne platform.

Tebay was convinced that if he built another ALL-like demonstrator—a technical project—it would be viewed as a toy by the war fighters, not as a weapons system. The ABL had to stand on its own merit by filling a role that no other weapons system could fill: shooting down ballistic missiles at the speed of light.

But that didn't mean that the technology capable of ensuring that ABL would succeed was ready for prime time. Far from it.

Tebay needed the Phillips Lab researchers to prove the physics; that is, to prove that a weapons-class laser beam could propagate hundreds of kilometers through a turbulent atmosphere and be held on a target long enough and accurately enough to destroy the target.

And that was no easy task.

Many argued—and some still do—that the technology to produce those desired weapons effects were not yet mature.[11] Enormous scientific problems had to be addressed and solved, and any one of them might cause the ABL program to fail.

For example, one early critic pointed out that an enemy could simply buff its missiles before launch, creating a highly reflective surface that would reflect the ABL laser and either prevent or lower the absorption of the high-energy laser beam. This might prove to be an easy yet effective way to counter the ABL.

The air force could not immediately prove using experimental data that this would not happen. As a result, the Phillips Lab researchers established a comprehensive laser lethality program that addressed this and other questions.

The program subsequently proved that there was little difference between the various highly reflective missile skins (both painted and unpainted), and hundreds of tests showed that little could be done to stop the absorption of laser radiation.[12]

This is because a metal missile skin is nowhere near the same highly polished quality of a mirror used to reflect laser light. These mirrors vary across their surface less than one-tenth the length of the laser wavelength—a tenth of a millionth of a meter. Compare this to the bumpy defects that can be felt on even a smoothly polished missile skin. Those defects cause varying properties of reflectivity on the missile skin, which ultimately results in absorption.

But this still did not mean that all Tebay had to do was assemble myriad technical parts and build the ABL. Scientific advances were needed to increase the efficiency of the laser, the point and track algorithms, and in numerous other areas.

Twelve years after Tebay established the SPO, and as the ABL prepares to make its first flight with the laser on board, some senior researchers point out that the ABL was still being constructed too close to the margin, and the science and technology base needed to be expanded to ensure success.[13]

In 1994, the ABL program office moved out from under the wings of the Phillips Lab and became an independent SPO (systems program office). ABL was then recognized by the USAF as an official weapons program.

Under pressure from Darlene Druyan, the senior acquisition civil servant in the air force, the ABL SPO was kept under 50 people, whereas the typical SPO ranged from 300 to 500 personnel.[14] This radical move allowed the ABL SPO to be agile, as Tebay wanted workers and not loafers, and it also kept the cost down.

But controversy was created when this small SPO flew against every other large acquisition program the air force conducted. Some felt that the ABL was being set up for failure. Unlike a typical new aircraft program that merely extended technology by taking an evolutionary approach, the physics underlying the ABL still needed to be proven, making this a high-risk, revolutionary venture. Some argued that the ABL SPO needed *more* people, not fewer, to ensure its success.

And in effect, ABL did just that, not by assigning additional scientists and engineers to the SPO but by purposely exploiting the scientific talent located quite literally next door at the Phillips Laboratory.

So in a way, the air force was conducting a shell game. Direct ABL costs were kept relatively low through using a small SPO cadre; at the same time, Phillips Lab, and subsequently the Air Force Research Laboratory's Directed Energy Directorate, researched the physics problems of beam propagation, laser material interaction, beam control, and pointing and tracking algorithms.

The air force toyed with the idea of physically moving the ABL SPO to Wright-Patterson AFB and placing the SPO under Aeronautical Systems Center (ASC), taking it from the Space and Missile Systems Center (SMC). This made sense in a command-and-control structure, as the ABL was actually an airplane and not a missile—and all airplanes built by the air force were under ASC.

However, the critical mass of scientific talent in high-energy lasers, atmospheric compensation, beam control, and other laser-related technologies resided at Kirtland, close to the Phillips Lab, and not at Wright-Patterson AFB. More importantly, at Kirtland the ABL SPO could continuously meet with the Phillips Lab researchers who were directly supporting their program. It was much easier to walk next door to get an answer than to travel from Ohio to Albuquerque.

Tebay made the decision to remain at Kirtland, and the generals allowed the fledgling ABL SPO to remain under control of SMC. When the SPO actually acquired an airplane, it would move under the control of ASC.

With that decision made, the ABL SPO was facing an intense time crunch to succeed. And to succeed, it needed to be fully funded.

At the time, the air force was confronted with competing funding requirements for its major acquisition programs, with its highest priority, the F-22 fighter, garnering the most attention and discretionary money. The ABL was considered the second highest priority air force acquisition program, but that did not make the situation any easier for Tebay. The SPO constantly had to defend its fledgling program against an onslaught of critics hungry for its funding, despite alleged air force support.

To make matters worse, laser system technology had to be developed during the design phase, in parallel with defining the mission and identifying the lethality mechanism—another ABL SPO anomaly, for in all other "normal" acquisition programs, their technology was already mature before their system was built.

The main criticisms facing the ABL were that atmospheric distortions due to turbulence could not be overcome; that these distortions would prevent the laser from propagating far enough to do damage; and that even if the first two problems were solved, the laser would not damage the missile sufficiently to destroy it.

So the SPO established parallel teams, heavily populated with Phillips Lab researchers, to address these criticisms and solve the problems associated with the fledgling laser technology:

- One team used the Fire Pond test range at MIT's Lincoln Laboratory to lay the groundwork for solving atmospheric propagation problems and controlling the beam's jitter.
- Another team conducted laser-beaming experiments to validate atmospheric compensation at White Sand's North Oscura Peak test range against test airplanes.
- Another team increased the beam's fluence, the amount of energy deposited on the target over a set time.
- Another team conducted lethality tests at Kirtland AFB and White Sands Missile Range to determine what was called the "lethality mechanism."

## The Lethality Mechanism

Critics kept asking how the ABL's laser would actually destroy a ballistic missile. Up to that time, the only hard data available on destroying air vehicles with lasers had been obtained by Project Delta, the navy's UNFTP, and the Airborne Laser Laboratory. But those involved drones and air-to-air missiles, not ballistic missiles, which were much larger and would be much farther away from the ABL than the other targets had been.

Would the ABL's laser heat the warhead, causing it to explode? It was easy to show using "back of the envelope" calculations that the ABL's laser energy would not be sufficient to do this with a heavily insulated warhead.

So how exactly would the ABL engage its target, and how would it kill it? Before the USAF went out and spent more than a billion dollars of taxpayers' money, this important question needed to be answered.

Researchers used the MIRACL (mid-infrared advanced chemical laser) at White Sands to demonstrate that the laser would not burn a hole in the side of a missile; rather, the laser energy would heat the missile and soften its skin. In flight, a missile's fuel tank pushes out against the missile's side with enormous pressure, creating what scientists call a hoop stress. The missile skin literally bulges because of the pressure exerted by the fuel tank.

The MIRACL experiment demonstrated that the ABL laser would soften the missile's side so much that the pressure in the fuel tank would cause the skin to burst, unzipping the metal around the missile's body. A highly publicized movie of this event was circulated in the press, and although some critics called the experiment a stunt, the demonstration left no doubt of the lethality mechanism.

As the lethality and physics questions were being answered, the ABL SPO pushed ahead with an industry competition. After initially awarding two, 2-year $20 million design studies to demonstrate the critical technologies, the contractors competed for a $800 million, five-year contract to propose the best technical solution for building an ABL.

Before awarding the contract, the SPO commissioned separate contractor teams to study airplanes that could carry out the ABL mission. Lockheed and Boeing independently studied planes that could carry 100,000-pound payloads, loiter at 40,000 feet for five hours, and be refueled in the air. The teams independently came up with the B-52 and the 747—the B-52 because its massive wings produced the lift needed to carry the laser's weight, and the 747 because of its immense cargo volume.

The results of this study were given to the industry as contractors competed for the ABL contract. And throughout it all, Tebay made it clear that the SPO did not want a demonstrator, but an actual war-fighting system that would be the first of a fleet of seven ABLs.

The winning team consisted of Boeing Missile Defense Systems, based in Washington, D.C., and Seattle, as the integrating contractor. Boeing had responsibility for the overall program management, as well as developing the battle management system, modifying the 747 platform, and the ground support systems.

As major subcontractors and Team ABL members, Lockheed Martin Space Systems, of Sunnyvale, California, was responsible for the target acquisition, beam control, and fire control systems, while TRW (later bought by Northrop Grumman Space Technology), Redondo Beach, California, was responsible for the high-energy laser.

In 1996, Colonel Mike Booen joined the ABL SPO as Tebay's deputy, and subsequently took over for Tebay when he retired in 1997 after 30 years of service to the air force.

Booen, now vice president for advanced missile defense at Raytheon, steered the ABL SPO through political turmoil and made many advances in technology before handing the SPO to Brigadier General Ellen Pawlakowski. Pawlakowski oversaw the transition of the ABL from the USAF to the Missile Defense Agency (MDA), formerly known as SDI (Strategic Defense Initiative).

The ABL was always considered the initial line of defense in a tiered, or multilayered, missile defense for the MDA. It was key to the so-called boost phase defense. The ballistic missile is most vulnerable as it is accelerating from the launch pad and is most visible, as the infrared signature from the exhaust plume is extremely bright.

The other two phases of missile defense—midcourse phase, when the missile is coasting high above the atmosphere, and the terminal phase, when the warhead is zeroing in on the target—are much harder to defeat, and are being addressed in separate MDA programs such as kinetic kill vehicles.

For example, in the midcourse phase the warhead is relatively cold in the sense that it is not accelerating, and it is difficult to spot, both visually and in the infrared. During the terminal phase, the warhead is accelerating toward the target until it reaches its aerodynamic terminal velocity. It has a very small cross-section (or visible area), and it is much harder to kill than simply unzipping the missile's fuel tank during the ascent phase.

Thus the ABL was moved to the Missile Defense Agency so the program could be fully integrated into the multitiered national defense missile defense program. This not only ensured that the target acquisition and command and control of the boost phase was integrated with the rest of the program, but people realized that no one system is perfect. There would be some "leakage," or missiles that would not be destroyed during the boost phase, and those missiles would have to be addressed in subsequent phases.

Coordinating the handoff of the missiles that escaped the ABL's boost-phase defense to the midcourse interceptors was crucial for identifying and honing in on warheads that were coasting high above the atmosphere.

Moving the ABL to the MDA was not accomplished without controversy. Several critics pointed to formidable technical problems remaining to be solved, such as ensuring adequate atmospheric propagation and ensuring the laser modules would generate enough power. The concern was that MDA would not invest enough science and technology into solving these problems, and the ABL might fail.

However, MDA has continued to invest in S&T, although not as much as some critics would like.

A more cynical view came from the perception that the air force was not really serious about fielding the ABL and was actually using it as a slush fund. This view purported that the ABL was moved to MDA because the air force was squirreling away cash for its number one priority—the F-22.

In any event, Colonel Tebay's deal with the generals to eventually move the ABL program under the Aeronautical Systems Center was OBE—overcome by events. In 2005, the actual test plane was stationed at Edwards AFB in anticipation of its first shoot-down, and General Pawlakowski's SPO still resided at Kirtland, although now reporting to the director of MDA.

So whatever the motivation for moving the ABL, the air force was now dependent on a defense agency and not its own service for ensuring that its antiballistic missile program was successful.

# 11

# ABL: The Airborne Laser

MILITARY PILOTS affectionately call it "zero dark early"—that time
in the morning before the sun rises when it would have made more sense to
stay up all night than get up so early. A cynical theory hypothesizes that military planes can't take off after 5:00 A.M.

On this morning vapor roils from fuel tanks, each filled with hydrogen peroxide, that dotted the desert runway jetting out from the Edwards Air Force
Base operations center. Green metal spheres 20 feet tall hold the poisonous
gas far away from the flight line, as any contact with water vapor would cause
the fuel to explode into a boiling inferno.

Engineers dressed in silver flame-retardant suits swarm around the tanks,
filling specially configured fuel trucks that transport the noxious liquid to the
waiting aircraft.

The place is the Mojave Desert, California, two hours north of Los Angeles,
home of every significant experimental aircraft flown, from Chuck Yeager's Bell

X-1 that first broke the sound barrier to the X-15 and Burt Rutan's Space-ShipOne, the first manned commercial foray into space.

Machines invented here are touched with destiny. If history holds true, these planes will fly faster, fly higher, or carry more weight than ever before. The philosophy here is not just to increase an airplane's capability by a few percentage points but to smash all records and advance the aerodynamic state of the art by leaps and bounds.

Today Edwards AFB, situated at the southeastern edge of the Mojave, is home to another revolutionary airframe, which is in some ways the most revolutionary of all: Air Combat Command's USAF YAL-747.

What makes this aircraft so different is not how fast or how high it can fly or how much weight it can carry but the cargo—the world's first airborne directed energy weapon: the airborne laser—ABL.

Dawn breaks as red shafts of sunlight stream down over the sandy desert floor. At the far end of the tar-lined runway, the YAL-747-400 freighter waits like a Sumo wrestler, wings drooping with loaded fuel tanks.

And like the Sumo, weight has always been a concern with the ABL. The specially modified 747 had been fitted with the world's largest titanium plate at Boeing's plant in Wichita, Kansas, for extra strength to hold the enormous laser modules. To accommodate the sophisticated COIL laser system, more modifications had been done to the 747 than any other airframe in history: in addition to the titanium plate, meter-wide holes were bored into the fuselage, and the entire nose section was replaced to make room for the beam control system. The weight tolerances were so tight that the ABL systems program office reviewed every pound of hardware brought on board the massive aircraft.

The ABL carries a full load of laser fuel—common household chemicals such as hydrogen peroxide, iodine, and chlorine. Once airborne, the chemicals will undergo several sophisticated reactions on board the plane that will transfer energy from a state of oxygen called "singlet delta" to iodine molecules, which in turn will emit laser radiation at a wavelength of precisely 1.32 microns—a little more than one-millionth of a meter in length, but long enough to avoid being absorbed or scattered by the atmosphere, and just the right length to slip through absorption resonances as it shoots from the laser to a target at the speed of light.

Throughout the first decade of the twenty-first century flight tests will continue as eventually a full squadron of seven 747s will be housed at the desolate California base. Edwards is far from civilization, the perfect place to locate the deadly mixture of noxious laser fuels and weapons-class beams that come with the ABL. Plus, its arid climate makes it ideal for preserving the delicate laser optics and beam control system.

Missile Defense Agency plans call for the ABL to be tasked as part of a multilayer missile defense, the first line of defense against enemy ballistic missiles. The ABL will target the missiles as they break through the cloud layer with their engines still burning, making them an easy target for the ABL's infrared sensors.

For rockets that manage to slip through the ABL's initial defense, the Missile Defense Agency is building fast-accelerating kinetic kill antimissile missiles, housed at the remote Kodiak Island launch facility in Alaska. The acquisition of the ABL was moved from the USAF to MDA so that this integrated multilayer missile defense would become a reality. And on paper, this is one of ABL's primary missions.

But other plans are being sculpted at the USAF Combat Air Command. Air force mission planners see the ABL as a robust, multitalented platform that can be used for more than stopping the threat of intercontinental missiles.

Forty thousand feet above the Pacific Ocean, the heavily modified YAL-747 airborne laser orbits in its combat oval racetrack pattern.[1] Since the plane can't hover like a helicopter or turn tight circles around a spot like an agile fighter jet, the mammoth ABL remains on station by flying in a nearly constant bank, which allows it to respond quickly to any threat.

Although stationed 100 miles off the coast, the ABL is guarded by a flight of F-15C air-to-air fighters, as is every other high-value asset such as AWACS (airborne warning and control systems), the lumbering command and control aircraft, and JSTARS (joint surveillance and target attack radar system), the GMTI (ground moving target indicator) platform that was rushed into service during the first Gulf War, just as the ABL is being rushed into service today, years before its IOC (initial operating capability) date.

The airborne laser is flying in combat-ready mode, its COIL (chemical oxygen-iodine laser) modules fully charged, laser fuel topped off, and the acquisition and tracking system fully engaged.

In this scenario the ABL has been called into position from its home at Edwards AFB to serve as the last line of defense. But instead of being deployed in a foreign war zone, this time the ABL's mission is protecting the homeland, 100 miles off San Francisco Bay.

Two days earlier, a cargo ship sailing from eastern Asia deviated slightly north from its charted course toward Los Angeles. This far from the coast, the small deviation would change its path from arriving at the Long Beach shipping port to aiming straight at San Francisco. Such navigation errors are common.

Since the cargo ship was still in international waters, normally the coast guard would contact the vessel when it reached the 12-mile limit of U.S. waters, and the ship would execute a slow, lumbering turn toward its scheduled port. But four weeks earlier the ship had been spotted leaving Iran, and with an elevated terrorist threat level, indications were the ship was carrying some sort of missile to be used against an American city.

In previous years the U.S. navy interdicted such ships on the high seas, aggressively combing them for contraband and weapons of mass destruction. But a series of embarrassing failures had made the navy gun-shy. In December 2002, the navy intercepted the North Korean freighter *So San* and discovered 15 SCUD missiles, identical to the medium-range rockets used by Saddam Hussein in the first Gulf War. But the supposedly illegal weapons were the legitimate property of Yemen, so in a highly publicized and embarrassing about-face, control of the vessel was returned to the captured crew and the navy backed off from its policy of intercepting ships in international waters.

The problem with this new, "enlightened" policy is that U.S. jurisdiction extends only 12 miles from the coast. If a missile-laden ship were to venture to within even twice that distance from a city such as Seattle, Los Angeles, or San Francisco on the West Coast, or Boston, New York, or Washington, D.C., on the East, and launch a missile carrying weapons of mass destruction, it would be too late for U.S. forces to react.

But the United States cannot afford to keep numerous fighter jets aloft in anticipation of this dangerous new threat. Fighters carrying air-to-ground missiles are limited by range—the jets must be within tens of miles from their moving target, and they cannot be launched unless there is a clear and imminent danger. In addition, it takes several seconds for the air-to-air missiles to reach their prey, and covert rockets carried by a seaborne platform might already be streaking off toward their target by the time fighter jets reacted.

It seems that terrorists may have stumbled on a key vulnerability, an asymmetric advantage of using a relatively cheap, slow-moving trawler carrying a hidden missile to slide up to a major American city. Employing several decoy trawlers would make it nearly impossible for U.S. inspectors to detect the real threat and respond in a timely fashion without turning international opinion upside down.

But the Air Combat Command planning staff refuses to limit the ABL to serving only as the first line of attack for ballistic missile defense (BMD). They believe that such a capable system should not be constrained. And in the face of an adversary intent on destroying our way of life, all assets should be employed. The only problem is that no one knows for sure what other missions the ABL is capable of performing.

As with any new weapons system, the real benefit is gained when the weapon is transitioned from the acquisition officials (the buyers) and turned over to the war fighters (the users). For its BMD mission, the ABL is equipped with state-of-the-art telescopes, radar, and laser trackers. There is no reason why the ABL cannot be used as a high-flying intelligence-gathering platform to peer hundreds of miles where ordinary airplanes are not able to see.

In addition, the ABL's weapons-class COIL laser is capable of destroying ballistic missiles over 100 kilometers away. Its baby brother—the COIL advanced concepts technology demonstrator, the advanced tactical laser (ATL)—is proving its capability to use much lower power levels, by a factor of 10 to 100 less than the ABL, to deter terrorists.

The enterprising war planners reason that the ABL could perform much the same function as the ATL, their own asymmetric response to the threat. Its highly resolved tracking can be used against slower targets than accelerating ballistic missiles, and its powerful laser beam can extend much farther than the distances needed to destroy intercontinental missiles to precisely deter a terrorist threat, thus using a strategic platform for tactical applications, except at distances vastly farther than what the ATL ACDT was designed for.

So instead of maintaining scores of fighter aircraft aloft in a chancy and leaky umbrella to provide protection against a seaborne threat, one ABL could accomplish the same mission not only faster and more efficiently but at the speed of light.

Flying at 40,000 feet ASL (above sea level), the ABL can remain combat ready for nearly seven hours. When it drops in altitude to refuel from a KC-135 tanker, it can extend its combat mode indefinitely. However, ABL crews limit their flights to a prescribed 12 hours unless an additional flight crew is carried.

The YAL-747 has been assigned the call sign "Laser One." With an elevated threat level and a current intelligence tip-off on several suspicious, slow-moving trawlers, the ABL left Edwards AFB earlier in the day to fly patrol.

Five hours after achieving "feet wet"—when the aircraft is flying over water—Laser One continues its monotonous mission, flying in a continuous bank as it covers the seas between Los Angeles and San Francisco. White contrails from high-flying F-15C air-to-air interceptors guarding the lumbering laser weapons system give comfort to the ABL crew.

Data from wide-area IR (infrared) remote sensors are fused with optical and radar imagery generated from the ABL sensor array. To track specific targets, the ABL uses an active laser ranger, a modified third-generation LANTIRN infrared seeker that uses a high-power $CO_2$ laser. The sensor acquires the target from the IRST (infrared seek tracker) sensor cue, tracks the target, and points the $CO_2$ laser for ranging. For antiballistic missile missions, this helps determine the missile's launch point and, along with trajectory information, where it will impact.

The active laser ranger skips from boat to boat, lingering on each ship steaming toward the West Coast. Sophisticated change-detection software analyzes each potential target, comparing its IR/optical/radar signature with imagery stored from the last time the sensors acquired it.

If there is no identical match, the entire history of images is transferred to trained battle analysts who pore over the data and ask key questions: Is there any unusual activity on board the vessel? Have any structures such as radar masts or tubes been erected or taken down?

Up to now, the changes in each vessel could be explained because of a different orientation of the ABL or the ship itself; or a ship may have jettisoned garbage or sewage into international waters. The crew of one ship had a barbecue on deck, which created a hot IR signature that immediately set off warnings inside the ABL. Even with a heightened terrorist alert, the war fighters slipped into the familiar role of hurry up and wait.

Suddenly a light flashes at the ABL battle manager's station. Specially qualified air force officers train for years to participate in combat on board the ABL. They're taught to assess the battlefield instantly and direct the pilot and onboard engineers in the complicated process of lasing a target.

The battle manager scans the display as information scrolls across a touch-sensitive screen. IR and visible images, along with threat parameters—target velocity, estimates of plume burnout, target size, and a threat analysis of the warhead—are relayed from the intelligence analysts.

The battle manager orders a close-up of the suspected target. Immediately a high resolution imager acquires the vessel and swings its telescopes for a closer view. The image of the ship's deck comes into view, canted at a weird angle because of ABL's banking turn and the rocking of the ship as it slips down ocean swells.

A smeared-out patch of hot metal glows eerily in the IR image. It looks as if someone had taken a blowtorch to the deck plate and heated up the metal. Another image shows ladar (laser radar) analysis of chemical elements, revealing a quickly dissipating cloud of hydrocarbons and noxious fumes. Whatever heated up the ship's deck is now gone, and the remaining evidence points to some sort of rocket fumes. But there is no indication of a rocket trail or heated columns of air that would point to a missile launch.

A warning flashes across the screen as the battle manager expands the acquisition sensor's angle of view. Instead of just focusing on the ship, the ABL's IRST sensor starts a methodical search pattern, looking for an IR signature that could be attributed to the flash. The battle manager queries the high-flying CAP (combat air patrol), but its side-looking and down-looking radar show nothing; the high resolution SAR (synthetic aperture radar) from the JSTARS ground moving target indicator is negative as well.

Suddenly picking up a weak signal from the IRST, the battle manager directs the tracking illuminator laser (TILL), a state-of-the-art diode-pumped, solid-state laser built to track the target, to the barely visible source. Although designed to accurately track a target point on a fast-moving ballistic missile hundreds of kilometers away, the same quick-response TILL sensor can sweep through a large area more quickly than current state-of-the-art sensors.

"Talley ho!" The assistant battle manager points excitedly at the shimmering IR screen, usually reserved for showing the portrait of an upward accelerating ballistic missile. But instead of showing a rising ballistic missile, the screen reveals the short, squat shape of a low-flying, stealthy cruise missile skimming just meters off the ocean surface.

The battle manager slaps at the touch-sensitive screen, calling up a sequence of algorithms that compute trajectories and positions backward in time. Using projected and captured data, within seconds, the screen shows the cruise missile was launched just minutes before from the deck of the suspicious trawler, confirming the origin of the target.

Another screen shows the projected trajectory and impact point of the cruise missile, assuming a straight-line approach and constant velocity: the wharf area of San Francisco, during the height of the tourist season.

Having previously received authority to engage a class of hostile targets, the battle manager issues a command to the acquisition manager. The ABL pilot automatically pulls the mammoth jet out of its racetrack pattern and aligns the craft onto an optimal path to engage the target. This target is vastly different from the typical ballistic missile the pilot and crew have trained to engage, but they follow a prescribed heading while bringing the ABL down in altitude so the laser will have less distance to travel in the optically thick air.

The ABL was optimally designed to engage ballistic missiles as they break through the cloud layer at 40,000 feet above sea level. At that distance the air is thin enough that the COIL laser will propagate efficiently through the atmosphere and not encounter the effects of atmospheric turbulence present at lower altitudes. But now, with the cruise missile skimming only meters from the top of the waves, the ABL needs to reach an optimal "sweet spot" in altitude that delicately balances the thickness of the atmosphere with the distance the laser will travel.

Simultaneously the singlet delta oxygen generators roar into life, as hundreds of gallons of BHP (basic hydrogen peroxide) pulse through the onboard plumbing. Once the singlet delta oxygen is generated, the excited molecule is sprayed through a supersonic nozzle as molecular iodine is introduced into the supersonic cavity. Mixing together in a roiling swirl of shock-heated chemicals, the excited oxygen transfers its energy to the iodine, shoving it into an "inverted state," ripe for a seed laser photon to cause the billions of quivering iodine states to release its energy—all in the identical direction, in the identical phase, and all at the speed of light, out of the cavity and into the optical train.

As the aircraft dives, the BILL (beacon illuminator) and TILL lasers rotate and lock onto the cruise missile. The BILL, a solid-state laser similar to the TILL, illuminates the target for the adaptive optics that correct the beam wavefront due to atmospheric aberrations.

Simultaneously the target acquisition computer churns through millions of possible engagement scenarios, performing a real-time parametric optimization of firing angles and altitudes, all being updated from ABL altitude, attitude, and velocity, as well as outside temperature, air density, humidity, and other data relayed back from the sensors.

Within seconds an optimal engagement solution is obtained and the firing sequence begins. Energy in the COIL cavity grows exponentially. Directed out of the cavity and bouncing across the optical bench, part of the beam is bled off and shunted away for diagnostics.

Wavefront information from the BILL laser, reflected off the cruise missile, is analyzed in hundredths of a second. The atmospheric distortion implanted on the BILL is quickly inverted and applied through a mathematical process called an FFT (fast Fourier transform) onto the 18-inch deformable mirror (DM), moved by 964 mechanical actuators located behind the mirror's surface.

The mechanical actuators are actually tiny pistons, each capable of responding in less than two-thousandths of a second, pushing their tiny piece of the mirror less than a millimeter. Imagine lying on a vibrating bed moved by a thousand tiny vibrators, pushing a few thousand times a second in a coordinated motion that exactly cancels your movements—except that this bed is a mirror made of highly reflective glass and the vibrators are so tightly orchestrated that they can exactly cancel aberrations caused by the atmosphere.

The target acquisition, tracking, and beam manipulation are ready. The COIL is engaged and an inferno of infrared laser radiation is unleashed.

Slicing through the air at 186,000 miles a second, laser energy boils surface layer after surface layer off the cruise missile's side, until the metal skin is breached and its fuel detonates from the tremendous heat. The cruise missile's high explosives warhead explodes in a fireball, as the rest of the missile plunges into the cold waters of the ocean.

Ferried by helicopters, special forces swoop down onto the trawler's deck to take down the terrorist crew. As the arrests are being made, the ABL climbs in altitude, turning its sensor suite to detect other threats possibly lobbed from innocent-looking ships.

......................

### Give a Pilot a Ball of Crap and a String and He'll Make a Game of It

The above scenario illustrates a typical nonstandard use that might be expected of the ABL. After all, you have a multibillion-dollar weapons system that has been fielded for a primary job of shooting down intercontinental ballistic missiles in the boost phase. Since that doesn't happen often, the United States will naturally use that national asset to solve other pressing problems.

Although the ABL's primary mission is for anti-ICBM, many predict it will experience the same innovative, unexpected use made of every new major technological device. Once an asset has been turned over to the war fighter, it's almost guaranteed that a new, perhaps even more important application will be realized.

History bears this out.

For example, take GPS. No one in their right mind was willing to pay billons of dollars for a device that would simply tell you where you were to an accuracy of centimeters, much less *when* you were to an accuracy of microseconds. For years people got along fine with AAA maps and standard navigation gear such as compasses.

Pilots used Loran, better known as radar navigation, to determine their location—which came in handy when they were flying in fog or clouds, and especially over the water.[2] But even then, no one needed more accurate navigation tools, since a pilot could always "eyeball" a landing once he flew in under the cloud layer.

Even more sophisticated navigational tracking devices were developed, such as star trackers, and then the ultimate, the INS (inertial navigation system), based on the accurate reading of the differences between gyroscopes. A laser gyroscope was invented that further refined the INS; so why in the world would anyone want anything more accurate than that?

This is precisely the criticism that a few, far-thinking scientists encountered when they proposed GPS. Undaunted, they argued that precise navigation would do away with the need for the sometimes inaccurate LORAN (then used for the majority of air navigation), and it could set a standard for everything from mapping to geo-location.

The air force scoffed at the idea. Why spend billions on a navigation system instead of a new class of fighters?

But luckily Congress agreed with the visionaries and the project was funded. And when GPS finally came online, guess what happened?

It fizzled.

At first nobody used it and nobody wanted it.

After all, why would you spend thousands of dollars buying a GPS receiver to find your location within a few hundred meters (for security purposes the air force diddled with the commercial GPS algorithm, making it from 10 to 100 times less accurate than the military version)? Even the military couldn't see a need to spend the extra money to obtain a new, uncertain capability. INS worked just fine, thank-you.

And then came the first Gulf War: the battle to liberate Kuwait.

Suddenly hundreds of thousands of soldiers and airmen discovered that AAA maps didn't work in the desert. Even worse, NIMA (National Imaging and Mapping Agency) didn't have accurate maps of Iraq because there weren't any recognizable landmarks for thousands of square miles.

The soldiers were frustrated, and before the war started they wrote home. As a result, moms rushed to sporting goods stores and bought up thousand-dollar GPS units and shipped them to their sons and daughters to help them find their way in the desert.

And it worked. Now everyone wanted GPS. The military upgraded the civilian ability to use GPS, and the number of uses exploded.

The commercial industry discovered that by putting a GPS on a golf cart, not only could the golf marshal keep track of golfers' location but management could optimize the flow-through of golfers by knowing the exact location (within meters) of a golf cart that was equipped with GPS.

And even better, a GPS equipped with a preprogrammed chip could serve as a virtual caddy: "Hole three is a dogleg, and at this spot, I'd use a four iron."

Who would have thought that GPS would be used by mountain bikers? And rental car agencies? Or even installed in kids' wristwatches so that moms can track their little ones?

GPS has engrained itself in society and has become indispensable to commercial navigation.

The point is that no one knew exactly what benefits would arise from investing in GPS, and we're only now just realizing the potential of this critical investment.

People are initially skeptical of new technology, but later they can't get enough of it.

This mind-set is not unique.

More recently, fighter pilots scoffed at precision-guided weapons: "Why do I need to be so accurate when a grease pencil on the cockpit windshield has worked for years?" Even the airborne warning and control system (AWACS) was decried for its centralized control of the air battlefield: the Soviets tried this and lost!

Many think the ABL will show the same resilience as other national systems and provide the United States with a capability that we haven't even begun to imagine.

After all, you've got a highly capable national asset. You put it in the hands of a war fighter and place that war fighter in a new, life-threatening situation. Do you really think this highly trained professional is going to freeze up and not function when faced with a novel situation? Our war fighters are taught to innovate, to think on their feet.

And just like that pilot with a ball of crap and a string, what else are they going to do?

## Invention to Innovation

The time that elapses between the invention of a weapon to the time someone finds a killer application—a use no one can live without—is known as the period from "invention to innovation." For example, although precision weapons were introduced in the 1960s in the Vietnam War, laser designators were not widely embraced until the news media televised scenes of accurate air-to-ground missiles shooting through windows in the first Gulf War. In that case the time from invention to innovation was roughly thirty years.

Every discovery goes through this period of "invention to innovation," sometimes referred to as the S-curve of technology development.

This is especially true of advances in the basic sciences, such as physics, chemistry, biology, and applied mathematics. In our case, the invention of the laser or high-power microwaves did not directly contribute to the mining, construction, manufacturing, or the defense industry. However, advances in directed energy technologies should profoundly affect these areas in ways no one dreamed when the research was being done.

Rarely did an invention result in an immediate appreciation for what it had to offer. Many today still wail at the perceived worthlessness generated by researchers.

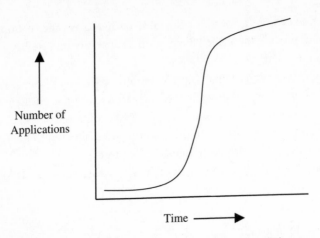

Figure 11.1    The S-curve of technology. At first, there is little apparent use or application (much less a killer app) for a discovery or invention; but in time, an accelerated interest occurs that flattens out.

But consider the recent history of the basic sciences.[3] Within a few years of 1875, discoveries that had been made since the 1600s—an incubation process lasting nearly 270 years—started to feed into the Western industrial technology base.[4] Is there anyone today with the patience to wait for the innovative use of an invention made 270 years ago? Over that period, the basic sciences laid the groundwork for explaining the basis behind the natural sciences, chemistry, and physics. These developments culminated in the development of rigorous engineering procedures responsible for the rapid evolution of technology. For example, the following list highlights a few of the hundreds of significant breakthroughs in the basic sciences in the mid-nineteenth century:[5]

*1850s:*
- Clausius's second law of thermodynamics
- Riemann's non-Euclidian geometry
- Maxwell's kinetic theory of gases

*1860s:*

- Dmitry Mendeleyev's periodic table of the elements
- Kirchoff's blackbody radiation
- Maxwell's electromagnetic equations

*1870s:*

- Louis Pasteur's work in food spoilage
- Van der Waal's gas laws
- Gibb's chemical thermodynamics

The innovative application of these discoveries was not immediately apparent; however, from Maxwell's equations sprang the basis for radio, TV, electronics, and computers; Dmitry Mendeleyev's work on the periodic table established the basis for modern chemistry; Louis Pasteur's work in food spoilage resulted in the science of bacteriology and modern biology; and we've already seen how the foundation for directed energy was similarly established.

On the surface, there seems to be a direct connection between discovery and application. That is, looking at the past reveals a simple path from creative spark to world-changing technology.

But the route from these scientific discoveries to the technology used in industry was long and circuitous, rarely linear, and never straightforward. One discovery begets another; a new application yields a wellspring of others. Rarely did an application leap directly from the mind of the inventor. Rather, it waited to be revealed by the technologist as an onion's inner core is peeled away, layer by layer.

The labors of research do not quickly bear fruit; the applications of basic research are measured in decades, not days.

For example, the time between invention and innovation for the fluorescent lamp took 79 years, the gyrocompass 56 years, the cotton picker 53 years, the zipper 27 years, the jet engine 14 years, radar 13 years, the safety razor 9 years, and the wireless telephone 8 years.[6] So although this long time-scale is a drawback of long-range research, their applications have proven that they can change the direction of society.

The same holds true for the application of lasers and high-power microwaves.

So although the ABL was built for a specific purpose, the innovative use of it still waits to be discovered. And it will take the war fighters to do just that—just like that game those pilots invented with a ball of crap and a string.

## 40,000 Feet over the Arabian Sea

Intelligence updates fed into the orbiting ABL show the location of a terrorist camp deep inside Iran. The camp contains a cache of high explosives—plastique, mortars, ground-to-air Stinger missiles, guns, ammunition, explosive firing caps, TNT, and chemicals—being stored for the fabrication of weapons of mass destruction.

The ABL, deployed less than 24 hours earlier from its base at Edwards AFB, is now flying a narrow orbit at 40,000 feet above sea level over the Arabian Sea. It doesn't have room to stray—to the north is Iran, and already the ABL's sensors have picked up Iranian long-range radar, painting the craft as it prepares to send out fighters; to the south is Saudia Arabia, an increasingly nervous partner in the continuing U.S. presence in the Middle East.

The ABL is constrained to fly above the international waters of the Arabian Sea. It dare not venture into Iranian airspace, and yet the terrorist weapons cache is several hundred kilometers beyond the ABL's ability to reach out and destroy it, as it did the cruise missile threat off the West Coast of the United States.

In this case, the terrorist camp is below the horizon, beyond the reach of ABL's supersensitive sensor suite. And even if the ABL's sensors could detect the storage cache, the atmosphere the beam would have to travel through is much too optically thick for it to work. Its powerful COIL laser would have to penetrate distances far beyond the beam's lethal range.

Undaunted, the ABL reaches its station and assumes combat duties. The lumbering craft flies its monotonous racetrack mission, protected by F-22

Raptors flying CAP, as the military's technological wizards prepare to unleash another ace in the hole.

Within the hour a huge, unmanned dirigible flies over 90,000 feet above the surface of the earth and positions itself just outside Iranian airspace. The lighter-than-air craft, a high altitude airship (HAA), is above the range of antiaircraft weapons and is outside the limit of ground-based SAMs based inside Iran's borders. Suspended by the HAA is a binocular mirror consisting of a receiving telescope and a transmitting telescope that can first capture and then redirect a high-energy laser beam to a target hundreds of miles away.

Seconds after the dirigible arrives on station, sensors on the HAA lock onto the ABL's TILL laser, establishing a tight lock on both platforms' position. Simultaneously, long-range sensors on the HAA geo-locate the terrorist storage facility, using coordinates forwarded by the ABL fire-and-control system. It trains its transmitting telescope onto the camp, establishes a second lock, and relays its status to the ABL.

Satisfied that it is now possible to "bounce" its high-energy COIL off the HAA, the ABL initiates its firing sequence, and laser energy is directed at the HAA's receiving telescope.

A fraction of a second later, the HAA is flooded with the COIL radiation. In a sophisticated onboard beam conditioning system similar to that of the ABL, the HAA corrects atmospheric aberrations in the laser imprinted from the beam's journey from the ABL to the HAA. Once corrected, the beam is conditioned for the second part of its journey, from the HAA to the ground.

Following the lower-power laser used for adaptive optics, the powerful COIL smashes into the terrorist encampment, heating the explosives, and detonating the cache of weapons.

As the ABL's beam shuts off, the HAA turns and chugs away from the border, its mission complete. The terrorists' supply of weapons has been destroyed without an incursion into Iranian territory—except for a laser beam traveling at the speed of light.

• • • • • • • • • • • • • • • • • • • •

Science fiction?

Well, which part? The ABL? No. Soon it's due to start testing against tactical ballistic missiles. Remember the ALL shot down antiaircraft missiles in the 1980s, so shooting a weapons-class laser from an airplane is a fact and not science fiction.

The high altitude airship? Again, no. High altitude balloons routinely fly at heights over 100,000 feet, and there are plans to build such platforms for military use.[7]

Well, what about the relay mirror? Surely that's sci-fi.

As it turns out, not exactly.

The Missile Defense Agency (MDA), the successor to the Strategic Defense Initiative, plans to demonstrate a dirigible carrying a binocular relay mirror. In an ambitious project code-named ARMS (aerospace relay mirror system), MDA plans a demonstration of an operational relay mirror based on an airship to extend the range of lasers by hundreds of miles.

But still, these are just plans, aren't they? Bouncing lasers off mirrors in the lab is not the easiest task in the world; but bouncing them off an airborne platform requires superprecise pointing, aiming, and tracking. So in reality, is this another case of something that's possible to do but not probable? Like Dr. Forward's antigravity on earth?

No, it's not just probable that something like this can be done; it's a fact.

In addition to being redirected from a high-flying airship, a laser has been redirected to something even more spectacular: a relay mirror orbiting in space, fixed on a satellite racing around earth.

The next chapter covers this amazing feat, a low-key but groundbreaking 1990s experiment that some critics dismissed as a stunt: RME, the relay mirror experiment.

# 12

# Making High-Energy
# Lasers a Global Weapon

In THE EARLY 1990S, a crew of air force researchers deployed to the tropical island of Maui. Normally any government presence in a tropical paradise would be heavily scrutinized by zealous government auditors, but the researchers had an iron-clad alibi for their presence.

Years before, the military's Rome Laboratory had deployed an elementary atmospheric compensation experiment at the top of Mount Haleakala, altitude 3,058 meters, latitude 20.7 degrees N, and longitude 156.3 degrees W. The 10,033-foot mountain on Maui is renowned for its breathtaking sunrises, tourists streaking down its slopes on bicycles, and incredible views of an ancient volcano crater set against lush plant growth.

Rome Lab demonstrated the first practical application of adaptive optics—a far cry from the current AO system now on the ABL but a convincing demonstration that the debilitating effects of atmospheric turbulence could be overcome by the physical manipulation of optics.

Although Maui was a tourist haven for elite sun worshipers and hippies, it was comfortable with a growing high-tech sector. Throughout the 1990s, largely through the efforts of Senator Daniel Inouye, high technology exploded and Maui became home to the USAF Maui High Performance Computing Center (located in Kihei), and the ambitious Maui Space Surveillance Complex (MSSC, formerly known as AMOS), which still serves the astronomical community as the largest (3.67 meter main optic and a 941 mechanical actuator deformable mirror) optical adaptive telescope operating at this altitude.[1]

But the air force researchers descending on Maui did not have adaptive optics or sunbathing in mind. They were consumed with greater accomplishments. They had arrived to conduct a breathtaking demonstration of a technological leap forward—reflecting a laser beam off a low-flying satellite and having the satellite direct the reflected beam onto a target on earth.

Incredible. This was a real-world example of something that can be calculated in physics (through the use of Snell's law of reflection, his law of refraction through the atmosphere, and Newtonian physics in calculating the path of the satellite and where it needs to point the reflected laser beam) but is an engineering nightmare to accomplish.

Consider the problem. A satellite races 600 kilometers overhead, traveling 7.7 kilometers a second or approximately 17,000 miles an hour. To reflect a laser beam off such a quickly moving target, its angular velocity and the distance the laser has to propagate must be considered. At 17,000 miles an hour, the satellite will move 15 meters (about 49 feet) away from its position when the laser travels from the ground, 600 kilometers away. At that distance, to hit a mirror 1 meter square, roughly 3 feet by 3 feet, means that the tracking accuracy must be held at around a constant millionth of a radian per second, or 30 millionths of a degree every second. In other words, the laser beam must be steadily held to an angle less than the area of a dime when viewed 10 miles away.

That's tough but not impossible.[2]

Some astronomers may be unimpressed, since they routinely track objects with that accuracy; but, as Paul Harvey used to say, look at "the rest of the story."

Not only must a satellite be illuminated with this accuracy, but the satellite's mirror must lock on to a target on the earth, which, from the satellite's point of view, is racing along at 17,000 miles an hour! As the laser propagates into the atmosphere, it diffracts as it encounters changing atmospheric densities due to the height above ground, the makeup of the atmosphere at those altitudes, and even the wind velocities, humidity, and grit, sand, and pollutants in the air.

So the complexity of the problem has just increased.

It's not good enough anymore just to track and illuminate a satellite. Now the satellite itself has to track a target on the ground.

That's easy to calculate but nearly impossible to engineer.

But in the 1990s, these air force researchers not only proved they could do it, but they did it by using a laser on top of Mount Haleakala, reflecting the laser off a mirror on a satellite passing over Maui, and illuminating a target at the base of the mountain.

Reflecting lasers off satellites sounds like an interesting science experiment, but why on earth should this matter today, and why is it important to the ABL?

Recall the scenario described in Chapter 11—the terrorist explosives cache in Iran. The ABL was flying patrol high above the Arabian Sea, and the terrorist camp was out of sight of the ABL's most advanced sensors.

No matter how high the ABL can fly, if a target is beyond its horizon, then detecting enemy encampments on an air platform is impossible (with the exception of using a locating device such as over-the-horizon radar, which is not present on airplanes).[3] Once again, it appeared that the United States might have poured billions of dollars into fielding a technology—the ABL—with little to show.

But that was not the case. With the help of a high altitude airship (HAA) carrying a binocular mirror, the ABL vastly extended its range and in the scenario took out the terrorist weapons cache.

So by combining the power of an airborne laser with mirrors on a dirigible, the ABL leaped from affecting a battlefield a hundred kilometers in radius to an area several hundred kilometers across. And because the area is proportional to the square of the radius (remember $\pi r^2$), if an HAA can extend the ABL's range by twice as far, the ABL can now defend an area four times as large as it could without the HAA; and extending that range by a factor of four results in an area sixteen times as great. Thus by simply extending the ABL's range, the entire battle theater might now be covered.

The HAA mirrors could also extend the range of a ground-based laser, or perhaps that of the army's MTHEL (a tactical battlefield laser weapon discussed in the next chapter).[4]

When one key component was added, the battlefield laser went from being a point defense to being an area defense. Now *that* is a dramatic increase in military effectiveness.

But why stop there? If an airborne reflecting mirror could extend the range of a laser to cover an entire battlefield, wouldn't it be possible to take another technological leap and make the laser a truly strategic weapon and cover the entire world?

The answer is possible, and an advanced technology known as the space-based relay mirror might do just that.

The air force's RME (relay mirror experiment) proved that a laser system could track a fast-flying object in low-earth orbit, reflect laser energy off a mirror, and illuminate a target. This seeded the beginning of the ARMS (airborne relay mirror system) project, a waypoint before sending a relay mirror to space that could eventually direct a laser beam to any point on earth. This is profound. We're not talking about bouncing a low-power laser from a desolate mountaintop to a target positioned a few miles away.

The air force is actually considering implementing a high-tech version of one of its visions, Global Strike, and not by launching a B-2 bomber from a Missouri airfield, flying a mission that takes 30 hours halfway around the world, and striking a target with smart bombs.

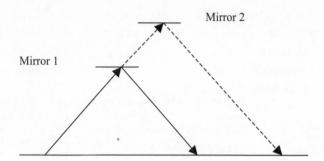

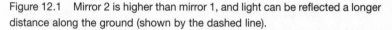

Figure 12.1    Mirror 2 is higher than mirror 1, and light can be reflected a longer distance along the ground (shown by the dashed line).

EAGLE (evolutionary advanced global laser experiment) envisions using a space-based relay mirror either to reflect a laser directly onto a target on earth or to bounce the beam off two relay mirrors so that the laser can engage the enemy around the world—and not just in hours or even minutes, but in milliseconds, at the speed of light.

Is this really possible? Can a laser beam be directed from earth onto a target around the world? Believing it can be done from a mountaintop is one thing; but clear around the world?

To understand how this might be possible, consider a short *gedanken* experiment. If you're outside and you sweep a laser pointer around, eventually you're going to hit a tree or a house, or some other person. Even if you were in the middle of the desert, because the earth is round, you can only shoot your laser so far into the distance before you reach the horizon.

Now take an optical-quality mirror and raise it above your head. Suddenly you can reflect that laser energy off the mirror farther than you could without using a mirror. In fact, the higher the mirror is positioned above the ground, the farther you can direct the beam (Figure 12.1). This is the principle behind the high altitude airship: the higher the airship is above the earth, the farther the distance the beam can be directed. We can extend this *gedanken* and raise the mirror to space: The higher we take the mirror, the farther the beam can reach.

Figure 12.2    No matter how high a mirror is above the earth (even if it were infinitely high above the ground), the longest distance light can be reflected is halfway around, to the other side.

This is exactly how the relay mirror experiment worked. The RME was positioned on a satellite around 300 miles above the earth when the laser was bounced from its mirror back onto the target in Maui.

So, it appears we've solved the problem! All we have to do is take the mirror into space, higher and higher, all the time covering more and more of the earth's area. In theory, if RME had been in a higher orbit, eventually we could have beamed the laser all around the earth and covered the entire globe, right?

Well, not quite. We reach a point, no matter how high we get, where the beam can no longer cover more than half the globe (Figure 12.2). Why? Because we've reached the horizon.

Because lasers travel in a line of sight, the beams can't wrap around the earth. Even two mirrors won't do it; a minimum of three relay mirrors are needed to completely cover the earth at any one time (Figure 12.3).

In addition, instead of using just one aperture (a monocular, or one-optic system) on a satellite that would have to continuously position itself as the satellite flew in low-earth orbit, a more sophisticated two-aperture (binocular, or two-optic) system would perform even better.

With two optics, the first aperture could acquire the laser beam from the earth. Placing a "beacon" on the aperture would allow the ground or air-based laser to precisely track the relay mirror and position its beam as the satellite passed overhead.

The RME used a beacon for its monocular system and proved that even with relatively crude 1990 technology (relative to the acquisition

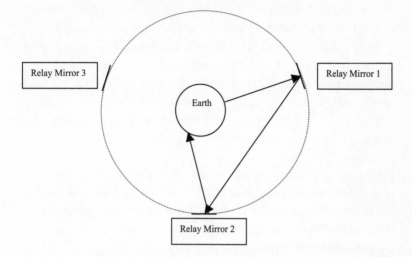

Figure 12.3    Three relay mirrors can reflect laser light to any point on earth. Since the laser beam propagates from mirror to mirror in the vacuum of space, the beam only widens due to diffraction and experiences negligible losses due to absorption or scattering, as a beam would undergo if it were reflected in the atmosphere.

and tracking technology that would be available in the 2010 time frame), it was possible to accurately reflect a laser to a ground target.

In a binocular system, instead of just reflecting the beam off an incident mirror, the beam would pass through the first aperture and be "conditioned" and cleaned up to remove any atmospheric distortions that might have occurred, even though the beam would have been first run through an adaptive optics system on earth. The beam could also be amplified, either by a chemical or solid-state amplifier section.

For example, work is now being performed at the Air Force Research Laboratory's Directed Energy Directorate on a gas-phase COIL-like laser that might enable the operation of a gas amplifier in space. A liquid-based system such as the present-day COIL on board the ABL cannot operate in space because in a zero-G environment the liquid would not work the same as on earth. Other lasers that may work in space include a solid-state system of fiber optic amplifiers or other gas-type systems.

Once conditioned and amplified, the beam would be redirected by the second aperture in a direction independent of the first aperture. A binocular relay mirror allows a satellite to track and lase a target for a much longer time compared to a monocular satellite. This is all due to the independent ability of the two apertures to acquire and target the beam.

Relay mirrors can dramatically increase the military effectiveness of a laser, either airborne or ground-based, by increasing the distance of the laser's reach. The success of ARMS will pave the way for deploying relay mirrors on a high-flying platform to extend a laser's lethality to cover the battlefield. And providing war fighters with the ability to deliver controlled amounts of laser energy in a precise location almost instantaneously will profoundly alter warfare.

Remember that scientific advances accelerated the growth of military effectiveness (Chapter 3). The atomic bomb was not truly considered a revolution in military affairs until it was combined with the 45-minute delivery time of an intercontinental ballistic missile, moving it from being detonated on a test stand in the desert or carried on a relatively slow-flying bomber. In the same way, the synergism gained by combining a laser and a relay mirror is another example of the accelerating increase in military effectiveness.

As revolutionary as laser weapons combined with airborne relay mirrors on a battlefield might be, however, they will be restricted to striking targets that are within the line of sight of the relay mirror's high-flying platform, on the order of a few hundred miles. So for targets deep within a major adversary's borders—say, a thousand miles—a relay mirror in space is the only viable solution for enabling a laser strike.

The alternative is launching a space-based laser (SBL), an ambitious project that was studied during the SDI era and more recently in the late 1990s. The studies concluded that it was indeed possible to field several

SBLs, but the cost approached several hundreds of billions of dollars. In addition, there was a technical risk in fielding huge mirrors (on the order of 15 feet) to reflect the megawatts of laser energy generated by a rocket engine, to precisely aim at targets located thousands of miles away. An SBL program would entail mobilizing nearly the entire industrial base to succeed.

Compared to an SBL, and especially in light of the successful RME program, the problems associated with fielding a relay mirror, although nontrivial, seem small.

Critics maintain that relay mirrors such as the RME are just a stunt, and that the real-world battle conditions a relay mirror would have to endure would be formidable.

Reflecting laser energy onto a target isn't easy—and it gets tougher the farther you want to bounce the laser.

But as with all weapon systems, once the capability is proven, more applications will be found. Relay mirrors will eventually move from high-flying platforms into space. Like other indispensable technologies such as GPS, AWACS, and even older technologies like radar that we take for granted, early success will ensure future war-fighter support.

# 13

# Other Major DE Weapons Programs

## MTHEL

ON AUGUST 21, 2003, the U.S. army and the Israeli Ministry of Defense announced the selection of Northrop Grumman Corporation's design for MTHEL, the mobile tactical high-energy laser, a prototype laser weapon capable of shooting down short-range rockets and artillery.[1] MTHEL is an advanced, mobile version of the THEL, an advanced concept technology demonstrator (ACTD) initiated in 1996 by the Defense Department and subsequently tested at the army's White Sands Missile Range in New Mexico.

THEL was designed and built by an international team led by Northrop Grumman, including Ball Aerospace and Brashear LP, with several Israeli companies, including Electro-Optic Industries, Israel Aircraft Industries, Yehud Industrial Zone, RAFAEL, and Tadiran.

In White Sands tests, THEL shot down 28 Katyusha rockets and five artillery projectiles. Now known as the MTHEL testbed, THEL demonstrated that it could shoot down the Katyushas fired singly or in salvos, providing a more realistic scenario of multiple rocket attacks such as might occur in combat.

Engineers have since extended THEL's capability by shooting down larger-caliber rockets with larger warheads. These rockets have twice the range and fly to more than three times the altitude as the Katyushas.[2]

Northrop Grumman has high hopes for MTHEL and sees the fledgling laser weapon system being fully integrated into the modern battlefield, just like any new traditional army system.

According to Northrop Grumman's MTHEL program manager, Joe Shwartz, "MTHEL will bring speed-of-light defense to the battlefield, but it will act and feel like any other air defense system. It will be operated by soldiers and supported in the field, mostly by the use of existing maintenance and logistical infrastructure. This enables both a seamless integration into current war-fighting concept of operations, while at the same time positioning the Army for the future."[3]

Northrop Grumman claims the cost per shot, primarily in the cost of the chemicals used to fuel the DF laser, is expected to be thousands of dollars—less expensive than the kinetic energy kill vehicle (KEKV) defense systems, in which a sophisticated rocket or projectile collides with a target to destroy it.[4]

Furthermore, since "traditional" KEKVs must impact a rocket or artillery shell, they are destroyed along with their target and are not reusable—a demonstration of the maximum capability of non-DE systems.

Unlike the USAF's ABL, which uses a COIL laser to weaken the metal skin surrounding the missile's fuel tank and allows the tank to erupt under its own internal pressure, MTHEL's DF laser heats the rocket's warhead. The deposited energy builds up and exceeds the detonation temperature of the high explosive, thus destroying the rocket in midair.

## ATL

In the late 1990s, the Boeing Company approached Dr. Hans Mark, then director of defense research and engineering at the Pentagon, about a proprietary reusable version of a COIL laser.

Although Boeing's COIL was relatively low in power compared to the strategic weapons-class ABL COIL, the reusable nature of the design intrigued Mark. Boeing's reusable COIL could be used for smaller, tactical missions to supplement, rather than compete with, the much larger ABL COIL.

As a result, the Department of Defense initiated an advanced concepts technology demonstrator using the tactical-level reusable COIL. The ACTD was called an advanced tactical laser (ATL). Special Operations Command sponsored the ATL ACTD, and a program office was established in Tampa, Florida, at special operations headquarters.

Because of the sensitive nature of special operations missions, there is little unclassified information available about the ATL. However, it is known that the laser will be carried on an airborne platform and will be evaluated for various tactical missions.

Lieutenant Colonel Brian Jonasen, the air force officer running the ATL program, worked with his public affairs officials to release the photograph on the following page (Figure 13.1).

### LAIRCM (Large Aircraft Infrared Countermeasures)

Shoulder-launched surface-to-air missiles (SAMs) have proliferated in recent years throughout the world. Although the United States lost no aircraft to SAMs in the first Gulf War, two aircraft (an F-16 and an F-117) were downed in Kosovo to SAMs in the mid-1990s.[5]

In May 2002, a U.S. plane was fired on by a SAM as it left Prince Sultan airbase in Saudi Arabia. On November 28, 2002, an Israeli Arkia 757 passenger plane was fired on while taking off from Mombassa, Kenya, most probably by a Russian-made SA-7 Grail, the most common SAM on the street.

The 1.4-meter-long SA-7 weighs 10 kilograms, has a 5-kilometer range, and carries a 1.8-kilogram high-explosive warhead. And of the 5,000 SA-7s estimated to be available to third world countries since 1966, the majority have been licensed to be built in Bulgaria, the Czech

Figure 13.1  The advanced tactical laser (ATL) advanced concept technology demonstration (ACTD) will evaluate a high-energy laser gunship capability on a C-130 aircraft to support special operations missions, including ultraprecision strike and operations. A tactical high-energy laser has the advantage of being able to produce both lethal and nonlethal effects on the battlefield and in urban operations. It can intentionally destroy, damage, or simply disable targets with no collateral damage. The advanced tactical laser demonstration represents a transformational weapons capability for the U.S. military. (Text and picture courtesy ATL program office)

Republic, Poland, Romania, Slovakia, and Yugoslavia (Serbia and Montenegro), and have been reverse engineered by Egypt, North Korea, and the People's Republic of China.[6]

In a one-month period from November to December 2003, four aircraft (a CH-46 Chinook helicopter, a Black Hawk helicopter, a DHL cargo plane, and a C-17 cargo jet) were all hit by SAMs in Iraq.

SAMs are typically heat-seeking missiles that hone in on the infrared (IR) signature emitted by aircraft engines. Thus early methods to counter the IR seeker involved creating an intense heat source in the form of high-temperature flares, which the aircraft ejected to fool the IR sensor.

Realizing that it was fairly easy to spoof their early "dumb" sensors, the builders of these surface-to-air missiles embedded complex seeking and tracking algorithms into their designs, and it became increasingly difficult for legacy anti-SAM devices such as flares to work; one can imagine a level of sophistication being embedded in the heat sensors that only looked for specific temperature distributions emitted from aircraft engines, rather than the typical blackbody temperature curve that a high-intensity flare might emit.

A cat-and-mouse game ensued with the SAM designers and the anti-SAM protection crowd, pitting advances in heat-seeking algorithms against flares and other devices specifically tailored to mimic the temperature distribution of engines. Even worse for the anti-SAM faction, the decoys merely lured the SAMs away from the aircraft as it tried to fool or spoof the SAM into believing that it was actually chasing an aircraft. It takes time for the heat sensor to lock on to a decoy, and all the time the SAM is rocketing through the air toward its target. And if that lock is ever broken, the SAM might reacquire the original target and end up taking down an aircraft. This is not a good situation.

There seemed no end in sight to the SAM/anti-SAM spiral until a solution was realized with lasers, one that could be executed at the speed of light.

The killer app of SAM defenses emerged when the dazzling power of lasers was combined with advances in pointing-and-tracking technology.

The LAIRCM (large aircraft infrared countermeasures) program gives transport and tanker aircraft the ability to defend against SAMs by using a laser that either emits or is modulated in several wavelengths in the mid-IR (called a multiband laser). This multiband, solid-state laser is integrated with a pointer and tracker turret that is typically installed on the rear of a large aircraft. The pointer/tracker automatically detects a SAM, tracks the missile's trajectory, points the laser, and holds the laser on the target while its multiband energy dazzles the SAM's guidance system.[7]

Initial production of the LAIRCM system for selected C-17 and C-130 aircraft was approved August 22, 2002, for a 2004 delivery to Air Mobility Command, providing the world's first defensive laser system employed on aircraft.[8] The $287 million program, scheduled to be completed in fiscal year 2009, is manufactured by Northrop Grumman and consists of an air-cooled, 10-pound mid-IR multiband laser that is compatible with the existing AN/AAQ-24(V) Nemesis directional infrared countermeasure system, currently carried by over 300 aircraft.[9] In addition, the Department of Homeland Security is considering the LAIRCM system as one of three candidates to protect commercial airliners against terrorist SAMs.

Although the LAIRCM system has performed successfully in tests, a large aircraft may have up to eight engines dispersed a hundred feet from each other, and there is a possibility that a jammed SAM might reacquire an engine if it flies too close to the aircraft. Multiple LAIRCM systems on a large aircraft such as a C-17 might prevent this from occurring, but no system is ever 100 percent foolproof. However, LAIRCM appears to be about as good a defense system as you can get.

## ZEUS

The scene looks like something out of a science fiction movie. A rugged Humvee bounces across the desert, topped by a laser mounted next to an array of RF antennas.[10] Radio-linked troops in combat gear fan out behind the Humvee, careful to step where the vehicle cleared their path of buried explosives only moments before.

Every few hundred yards the Humvee spins to a stop. Onboard electronic detection equipment receives pings from buried explosives, geo-locating the deadly IEDs (improvised explosive devices), and the solid-state laser cannon swings into position. Inside the bowels of the Humvee, diode-pumped flashlamps energize an array of neodymium-glass discs, all arranged at critical angles to the beam and engraved with intricate patterns to carry the heat away from the laser.

An invisible, coherent beam of 1.06 microns emanates from the solid-state heat capacity laser with enough power to heat steel at 200 yards.[11] The laser heats the target's outer casing until the munition is destroyed through internal combustion—a slow burn—rather than being exploded. This kill mechanism differs from the THEL (which detonates the enemy's Katushya warhead) and thus allows a "safe" and remote way to neutralize land mines.

......................

Only a few years before, the solid-state heat capacity laser (SSHCL) was invented at the Lawrence Livermore National Laboratory, a national security lab located just east of San Francisco. Livermore researchers won a prestigious R&D 100 award for their invention, and the army is now funding the lab to produce a high-power version of the SSHCL that may someday reach 100 kilowatts. Replacing the diode-pumped flashlamps with diode lasers is just one step toward achieving this increase.

The army trumpets several possible missions for ZEUS, including:

- Clearing active and test training ranges of exposed land mines
- Clearing airport tarmacs and runways
- Clearing unexploded ordinance from battlefields or for peacekeeping missions
- Humanitarian clearing of land mines[12]

ZEUS was deployed on March 18, 2003, in a combat zone, becoming the first high-power laser system to be used on the battlefield. ZEUS was used in Afghanistan for six months in support of Operation Enduring Freedom, where it successfully neutralized over 200 pieces of unexploded ordnance and negated 51 targets in a 100-minute period. In the months of testing before being deployed, ZEUS destroyed over 1,400 targets and completed its operation in a variety of environments, including rain and extreme heat.[13]

However, despite ZEUS's success, the system is limited to neutralizing targets that are not deeply buried. The system uses its laser to remove small amounts of dirt by heating the soil and "popping" the dirt away like popcorn; but this is only practical for shallow-buried targets. Because the war fighters must know the location of the target, ZEUS must be used with a detection technology to accurately locate the munition.

But even with these limitations, war fighters now have a capability to stand off nearly 300 meters and neutralize unexploded ordnance. This greatly reduces the risk for those charged with conducting operations in a battlefield, as well as aiding to clear mines from civilian areas.

# 14

# The Day After Tomorrow

## Will High-Energy Lasers or HPM
## Cause a Revolution in Military Affairs?

ALTHOUGH RADIO FREQUENCY (or microwave) weapons were designed as early as World War II, years before the first laser was invented, research investments in high-energy lasers have dwarfed that of high-power microwaves by almost a factor of 10 to 1.[1] The reason is that lasers can propagate weapons effects much farther than microwaves. However, weapons effects occur at very high power levels, and it has taken over four and a half decades (from 1960 to 2005) for lasers to achieve weapons-class powers, typically defined as in the megawatt range.

In Chapter 3 we saw that merely building a high-power laser did not a weapon make. Beam control, target acquisition, tracking, beam stability, and beam shaping are all important aspects of a laser weapon. These disparate parts must be integrated to make a weapons system.

However, integrating these components is still not good enough to make a robust weapons system. A robust, high-power laser weapons system cannot add an asymmetric advantage unless it can be used, and to use a laser, the beam must reach its target.

The reason is that lasers propagate in line of sight. With a powerful enough laser beam, if you can see a target, you can kill it. Simple enough.

But the problem is in seeing it. For example, if a target is over the horizon and a relay mirror is not available, then a laser can't kill it. And if it can't, then it's not much use to the war fighter.

The same argument was used with the hydrogen bomb.

The United States exploded the first H-bomb in the 1950s, and the event was hailed as a stunning achievement that vaulted the United States over the Soviet Union in the growing arms race. This was a scientific coup and demonstrated that the physics predicted by the Ulam-Teller secret was no longer a theory but was now a fact: H-bombs work.

But merely exploding the H-bomb did not give the United States an asymmetric advantage over the Soviet Union. In other words, it did immediately change the nature of warfare. It might have made a difference at the negotiating table or boosted national pride and prestige, but at the time the United States did not gain a strategic advantage.

The reason is that in the 1950s, the H-bomb was a multi-ton device that had to be assembled in place because it was too heavy, too unwieldy, and far too fragile to be carried on an airplane. An aircraft carrier might have worked, but even then, the H-bomb was more of a science experiment than a true strategic weapon. In that time period the H-bomb would not have hurt the Soviets unless they flew over the experimental site in the South Pacific.

So what did make the H-bomb an asymmetric weapon, a revolution in military affairs that would change the very nature of the way nations waged war? It was the technology that allowed the United States to shrink the device to a "small," self-contained warhead on the order of 2,000 pounds and integrate that warhead on an ICBM.

With that technology, the United States leaped from conducting an interesting high-energy density physics experiment confined to a small island in the South Pacific, to having the capability of delivering several million tons of explosive power with approximately 100 to 1,000 times greater destructive energy than the Hiroshima bomb—and deliver it anywhere in the world within 45 minutes.

Now *that* is a leap in military capability and a true revolution in military affairs. It was the integration of the H-bomb onto an ICBM with miniaturized weapons engineering that resulted in the asymmetric advantage, not the initial physics experiment.

In the same way, the United States produced megawatts of laser power—enough for weapons applications—in the early 1970s. But this did not produce an asymmetric military advantage. The early megawatt laser systems needed some way to transport their destructive energy to a target, to see the enemy.

## Taking the Beam to the Target

One way to take the beam to the target is to transport the laser by air. And as expected, the air force's airborne laser (ABL) is built to do exactly that.

The ABL uses a weapons-class laser system, but it has to be carried in the most heavily modified 747 ever built. The present result is a platform that is loaded to the max, with little margin to add additional weight, say laser fuel, to lase longer, since adding weight cuts down on the time the ABL can stay in the air. Adding a refueling capability will extend the time the ABL can spend in combat, but that extra time also stresses the crew and airframe.

The system will surely evolve, and more efficient laser systems will be introduced onto the ABL platform, but this process will take years. So for the foreseeable future, although the ABL will play a key role in certain combat situations—some that have not yet been envisioned—there is a limit to what the ABL can do.

This is not a criticism but an observation that is true of every weapons system deployed in combat. The warriors work with what they're given and the ABL will serve superbly, operating within the range of its lasers and sensors.

Advanced tactical laser (ATL), the shorter-range, tactical airborne COIL laser system, began as an engineering feat looking for an application. Someday it may prove to be a key part of special operations warfare. And again, it will evolve, in both technical capability and application.

Here the same observation holds true. The ATL will have a capability defined by laser range and the flexibility of the platform that transports it. It will deal with a class of tactical targets, but there are threats it was not designed to handle. It will complement the ABL, but these two weapons systems are not a panacea, a silver bullet for all threats (e.g., artillery or mortar attacks on ground troops).

The army's MTHEL will be deployed to protect against Katyusha-class rockets (and could have an ancillary mission to protect against some heavy artillery), but it takes days to transport, set up, and nurture a long logistics trail ranging from chemicals to specialists to make the system work. The army has high hopes for MTHEL, and, as with the ABL and ATL, the system will evolve. But it isn't the final answer.

With these examples, almost everyone would agree that ABL, MTHEL, and ATL aren't the end of the road for directed energy.

## Why the Future Looks Bleak for High-Power Microwaves

High-power microwaves in the form of long-wavelength radars achieved weapons-class power levels decades ago. But reducing the size of the system to something smaller than a battleship has kept the technology off the battlefield for years—until the Active Denial effect was discovered and exploited.[2]

It's one thing to use a relatively low-power Active Denial system against humans. Fielding an ultra-high-power HPM system against electronic equipment, the Holy Grail of researchers for years, is another. Really high-power HPM systems, the kind needed on a battlefield, are presently unfeasible.

A senior government official once said, "The smarter the weapon, the dumber HPM can make it."[3] This may be true sometime in the future, but it takes enormous power to do that, and HPM has two things going against it.

First, at very high power levels, microwave energy creates a plasma in air that prevents the microwaves from propagating.[4] This is called the atmospheric breakdown limit.

A simplified explanation is that when the microwave energy is greater than the energy binding the air molecules together, the air molecules dissociate into a hot, gaslike state called a plasma. This is not the kind of plasma found in blood. Rather, it is the fourth state of matter, after solid, liquid, and gas. Most solid matter turns into a liquid when heated (although some matter sublimates directly into a gas); when heated more, that liquid turns into a gas; and when heated even more, electrons are ripped from the gas molecules and a hot, soupy state of matter called a plasma is formed.

Plasma makes up over 99 percent of the universe. It is the stuff that makes up stars, the solar wind, and interstellar gas.

And just as astronauts lose radio contact with ground control when a plasma is created around their spaceship as it enters the atmosphere, microwaves can't transmit through atmospheric breakdown.

It's not good when your weapon can't propagate. It's like having a silver bullet that won't shoot farther than a few feet. It looks neat, but it isn't practical.

There is a second problem with HPM: because the wavelength is tens of thousands of times longer than a laser's wavelength, HPM spreads out 10,000 times more than lasers. Remember the discussion of diffraction in Chapter 5? Whereas a laser beam can shoot hundreds of kilometers, up to space and back, without spreading out much, an HPM beam isn't good for much more than a kilometer or so. Although it takes relatively little microwave energy to kill electronic devices, it's tough (if not impossible) to get the microwaves to the target.

Remember that the H-bomb didn't provide an asymmetric advantage until it was miniaturized and integrated onto an ICBM. In the same way, an HPM weapon isn't much good until you can figure out how to get its microwaves close to a target without spreading out, no matter how powerful it is.

And warriors don't like to get too close to their targets; as a rule, the nearer they are, the more danger they're in. The farther away from the enemy they can get and still wreak havoc the better.

Finally, the technology to shrink an HPM system to a size that can be reliably used on a battlefield is still decades away. Just as using fusion energy to solve the world's energy crisis is "just over the horizon" (and for some reason is always twenty years away), the same might be said for HPM.

The only attempt to exploit nonpersonnel uses of HPM is through a relatively new, very high-risk program that many view as a last-ditch effort to use anything to help with the IED (improvised explosive device) threat in Iraq. Although little information has been released on the navy's highly experimental NIRF (neutralizing improvised explosive devices with radio frequency), there are plans to deploy this device to Iraq.

NIRF only works at very close range and, according to the program manager Dr. Dave Stoudt, a cadre of researchers is needed to operate the device.

In conclusion, a decade or so from now, HPM weapons may be used on the battlefield to attack electronic targets; but until the problems of shrinking 30-foot-diameter antennas, miniaturizing megawatt power sources, and overcoming atmospheric breakdowns are solved, HPM will not be feasible. Only millimeter waves, and then with a limited range against humans in the form of Active Denial, appear plausible in the near future.

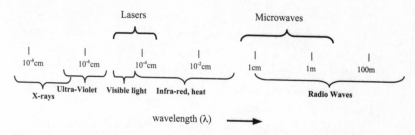

Figure 14.1  A sketch of the electromagnetic spectrum, with wavelength increasing from left to right. Lasers, on the left, have wavelengths 10,000 times smaller than microwaves, shown on the right.

But then again, history is always full of surprises. And technology leaps forward because of previously unimaginable breakthroughs.

## Lasers and High-Power Microwaves Are More Alike Than You Think

If the use of lasers and high-power microwaves is limited, what about other forms of directed energy? Are lasers and HPM all we can expect to exploit from the electromagnetic spectrum?

Recall that lasers and microwaves are just manifestations of the same thing, the electromagnetic spectrum. Lasers and HPM both consist of photons, or electromagnetic waves that have different wavelengths. Laser wavelengths run from ultraviolet to infrared—from 0.4 to 0.7 microns (or 0.16 millionths of an inch to 0.28 millionths of an inch), while high-power microwaves are generally defined as having wavelengths of anywhere from a meter to a centimeter (or 3 feet to a third of an inch). That covers just a small part of the electromagnetic (EM) spectrum. Figure 14.1 shows a part of the EM spectrum, with the wavelength increasing from left to right.

So what does that leave us? Let's take a look at what the rest of the electromagnetic spectrum has to offer, and why it might or might not be wise to exploit that part now.

First, there's the rather long part of the EM spectrum—long meaning the length of the electromagnetic wavelength, measured from crest to crest. The long part is generically known as the domain of radio waves. This covers wavelengths from a tenth of a centimeter (EHF, or extremely high frequency waves) down to waves over 100 kilometers in length (VLF, or very low frequency).[5]

The interaction of radio waves with matter is well known and has been documented for years. Recall from Chapter 5 that waves of the electromagnetic spectrum generally have to be the same size of the target or object to cause any damage. In a simplified view, lasers burrow into solid material quite well because their wavelengths are about the same size as molecules. Lasers can thus deposit their energy and "resonate" with the size of the solid material they hit, including metals.

On the other hand, although high-power microwaves can penetrate building walls and disrupt computers, they can't penetrate metals and don't do much damage to things like trucks or missiles. Instead, they interact with targets that are the same size of its wavelength (meters to millimeters), such as human skin and wires in electronics. This coupling, a measure of the amount of interaction, is greater for things that are the same size as an HPM wavelength.

This means that radio waves don't interact efficiently with targets unless they are the same size. And since radio waves are hundreds of meters to hundreds of kilometers long, they pass through most material and aren't much of a threat.

High-power microwave wavelengths are the longest part of the EM spectrum that can be used effectively as a weapon. And we've seen how difficult it is to produce an HPM weapon. So if we're going to exploit the EM spectrum, we've got to move to shorter wavelengths.

Most lasers work at only one wavelength unless the frequency is doubled or tripled through the use of nonlinear materials or exotic techniques. There is ongoing research in exploiting frequency overtones, much the same way sound overtones are obtained in audio devices such as pianos. But for the most part, laser wavelengths are fixed, and exploiting a frequency not presently available requires research that can take years to succeed.

Wouldn't it be nice to have a device that could lase at any frequency? And even better, suppose we weren't limited to lasing at normal laser wavelengths—suppose we could lase at *any* wavelength or frequency, as high as gamma rays or as low as microwaves, or anywhere in between?

Science fiction? No. Such a device already exists and has worked for decades, years before the first traditional laser was invented. It lases in a unique manner, unlike any other laser invented. It's a remarkable device with a remarkable story, and it's called the free electron laser (FEL).

## FEL: Free Electron Laser

We've seen how lasers can create intense, focused beams of light—electromagnetic radiation ranging from ultraviolet to the near infrared—that can be projected thousands of kilometers through space. And with the help of relay mirrors, those beams can be projected to any spot on earth, provided there is a direct line of sight from the mirror to the target.

We've also seen how laser beams propagate differently depending on their wavelength. Because of the diffractive nature of light, the shorter wavelengths diffract, or spread out, less than the longer wavelengths (see Chapter 5). This is true whether the laser propagates in space or in the atmosphere, as diffraction is a fundamental characteristic of all electromagnetic radiation.

In space, there is nothing to scatter or absorb laser radiation. A laser can travel in empty space forever, unimpeded in its journey to the edge of the universe, unless it is absorbed or scattered by interstellar dust or a stellar object. Or unless it journeys close to a massive gravitational body, such as a black hole, which actually curves space and causes the light to bend, much as an optical lens does.[6]

But in the earth's atmosphere, a laser beam is scattered and absorbed by air molecules, water vapor, or dust. Longer wavelengths scatter less than shorter wavelengths; our sky is blue because the shorter blue wavelengths of sunlight are scattered more than the longer wavelengths. And some wavelengths, such as gamma rays, are so highly absorbed that they can't propagate more than a few feet in air.

All wavelengths are scattered or absorbed, obviously some more than others. For example, infrared radiation created by $CO_2$ lasers has a wavelength of 10 microns (i.e., 10 millionths of a meter). This wavelength is strongly absorbed by water vapor, and $CO_2$ lasers do not propagate well near the ocean.

On the other hand, iodine lasers produce infrared radiation 1.315 microns in length, about seven times shorter than a $CO_2$ laser, and this wavelength is not absorbed nearly as much as $CO_2$ wavelengths in the atmosphere.

Since some laser wavelengths propagate through the atmosphere "better" than others (i.e., are scattered or absorbed less), what is the best wavelength for a directed energy weapon? Is there an optimal laser that could get around these debilitating effects? And could this laser somehow be "tuned" so that its wavelength could rapidly change to take optimal advantage of varying atmospheric conditions, such as thunderstorms, dust storms, or even volcanic activity?

In other words, is there a laser whose only drawback might be that it lases in a "diffraction-limited operation"—one that would not be absorbed or scattered by the atmosphere and would nearly operate as if it were in empty space, yet be powerful enough to be regarded as the Holy Grail of lasers?

Does such a laser exist? And if so, what is it?

Well, the answer is yes, but there is a heck of a lot of baggage that goes along with it.

Free electron lasers (FEL) represent a unique way of creating laser radiation without the use of chemicals, crystals, or any of the traditional means of generating beams. They are classed as electric lasers and can produce any wavelength, from extreme ultraviolet to microwaves.

FELs are so different from traditional lasers that they have more in common with high-power microwave sources than with high-energy lasers. And that difference is precisely how physicists stumbled onto this fascinating new class of lasers.

One of the reasons physicists work in their field is that they feel a child-like wonder at inventing new ways of doing things. They want to understand from the very beginning why things work the way they do.

Physicists don't take other people's word for it. They want to see for themselves exactly why something works. In a way, you can classify them as being from Missouri, the Show-Me state.

To some, this process of wanting to know why things work at a fundamental level is infuriating. After all, these interactions have already been derived, so why is there any need to verify them again?

The most obvious answer is that physicists are just a different type of people. (Their significant others will tell you that they are a different species altogether.)

Physicists want to understand things for themselves. Experiencing the reinvention is akin to inventing; by confirming the fundamental experiment, they are somehow there at the beginning.

So what does this have to do with directed energy?

A lot.

Transport yourself back to 1971, when coherent radiation in the form of lasers and microwaves was becoming so commonplace that most people thought that all the ways to produce this radiation had been invented. This was it. There's no place to go.

Chemical lasers promised megawatts of power, as the first 100-kilowatt laser had been demonstrated by Gerry just three years earlier at Avco. And generating microwaves using solid-state devices was all the rage.

One of the drawbacks of this era was that chemical lasers seemed limited to producing light (not actual visible light but infrared light roughly defined as from 10 μm to 0.1 μm). Because limitations in exciting energy levels obtain from the chemical bond strengths of chemical reactions, laser light of less than 0.1 μm would never be obtained.

Here is where the questioning nature of a physicist becomes important.

Not content to rely on the bond strengths of chemical reactions, supersonic expansion, or the energy stored in crystals such as ruby to produce the necessary energy levels needed for lasing, J. M. Madey turned sharply away from conventional thinking and proposed the first free electron laser.[7] Technology developed from his proposal is still vigorously pursued today.

An FEL is essentially an electric laser; the laser light is created by an accelerating electron beam as it "wiggles" back and forth through a series of alternating magnets (more on this later). The phenomenon was demonstrated in the 1950s using an alternating series of magnets known as a "wiggler" and a radio frequency linear accelerator.[8] As a result, the ubitron was invented in 1957, a precursor to the modern day FEL. The ubitron produced 150 kilowatts of microwave power, an amazing feat even for today's technology.[9]

Remember that a laser is defined as light amplified by the stimulated emission of radiation. So how is this light produced? Nowhere is it dictated that this light is produced by a chemical, gas, or solid state reaction. Light is light, no matter where it came from. So, reasons a physicist, if there is an appropriate medium so that the light can be amplified, then so much the better.

So let's consider another way method of generating light.

Recall that HPM sources were given a boost by the invention of the vircator (virtual oscillating cathode), which was developed by the high-energy charged particle community, specifically electron beam propagation experiments simultaneously performed by Russian and American scientists.[10] These experiments proved that converting the energy from high-power electron beams to high-power microwaves was possible. These experiments came out of a detailed theoretical understanding of the relationship between charged particle beam transport and microwaves fields.

In much the same manner, the development of traditional microwave tubes resulted from a detailed understanding of the interaction between RF (radio frequency, or microwave) fields and the transport of electron beams past various cavities. These beams were nonrelativistic; that is, their velocity did not reach a significant fraction of the speed of light, and much knowledge was gained in the 1950s and 1960s about their characteristics.

In contrast, the "traditional" high-energy accelerator community (e.g., at Fermilab and other places) was accelerating particle beams to relativistic velocities, where the beam's velocity approached a significant fraction of the speed of light.

It was discovered (as theoretically predicted) that when the trajectory of a relativistic beam was slightly curved, radiation was emitted from the beam. It was possible to accurately calculate the wavelength and direction of this radiation.

Those who create new lasers in some ways resemble the garage rock band that suddenly vaults into stardom or the unknown writer who makes the *New York Times* best-seller list overnight. People gape in amazement at the seemingly unearned stardom of the new pop sensation or the erudite novelist. Where did they come from? And most people think that they too, although they have little musical or writing ability, could emulate their success.

But what is hidden to the public is the hard work, the hours at the keyboards, going over and over that creative lick to make it perfect, or the solitary writer struggling to come up with just the right phrase.

Just as the success of a band, a novelist, or any creative pursuit depends on years of preparation, advances in science and technology occur the same way.

Nothing "just happens" in a vacuum. Discoveries are built on the efforts of others in the past. Isaac Newton recognized that he stood on the shoulders of giants.

But what is not fully appreciated is that the flash of inspiration, the *aha!* moment that defines a radically new advance, is not linear or preprogrammed. This is the stuff that results from insight and relies on the knowledge and work obtained by others.

So in 1960, when Theodore Maiman demonstrated the world's first optical laser by wrapping a coiled flash lamp around a rod of ruby crystal, he didn't act in a vacuum, without knowledge that the stimulated emission would happen—it was expected. And even more importantly, it was confirmed.

Maiman performed that experiment by building on years of groundwork laid by researchers in the past. He wasn't a member of the "band from nowhere" that suddenly created a megahit. On the contrary, the scientific basis for lasing had been established years before. We've already seen how Schwalow and Townes's work on the maser proved that stimulated emission could occur with microwaves, destroying the urban myth that Maiman's ruby laser was the world's first evidence of stimulated emission. Even more astounding, Maiman's laser wasn't even the first time that stimulated emission had been observed with visible light.

Coherent radiation from stimulated emission was first documented in 1953 by H. Motz and a team of collaborators using a radio frequency linear accelerator to drive an electron beam through a series of alternating magnets, a precursor to the ubitron.[11]

Alternating magnets are set up side-by-side with the north and south poles to create an alternating magnetic field. This arrangement is now known as a wiggler because a charged particle traversing this alternating field "wiggles" back and forth as it experiences the alternating forces (Figure 14.2).

This wiggling results in radiation, or photons being emitted that are all traveling in the same direction and with the same energy and phase as other emitted photons. In other words, these photons are coherent. This is because the electrons in the beam are coherent, and coherently wiggling electrons radiate coherently.

The coherent light is equivalent to that obtained from a "one pass amplifier" from a laser, and for all intents and purposes, Motz's team actually created the world's first maser in 1953 from an FEL. If Motz would have used magnets spaced closer together, a more energetic beam, and a higher-energy RF accelerator, he would have created the world's first laser.

In the years from 1957 to 1964, several experiments were conducted with wigglers spaced farther apart, and coherent radiation in the microwave region was obtained. This microwave equivalent of the laser, or

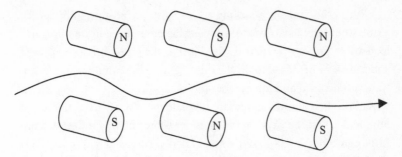

Figure 14.2    An electron beam wiggling through a series of alternating magnets. The electrons in the beam are coherent, which radiates coherent radiation.

maser (microwaves by amplification of stimulated emission of radiation), actually occurred years before the seminal Maiman experiments. For example, R. M. Phillips demonstrated a maser based on the free electron laser (called a ubitron, for undulating beam interaction) that produced over 150 watts of power at wavelengths of 5 millimeters.[12]

With help from congressional funding, in 1999 the Laser Processing Consortium was established to build an FEL called the IR-demo at the Jefferson Laboratory, located in Newport News, Virginia. The consortium consisted of a partnership of high-technology manufacturers, start-up companies, universities, the Commonwealth of Virginia, and the U.S. navy, and built a kilowatt CW (continuous wave, i.e., not pulsed) FEL.

The navy's interest in FELs comes from the need to field a weapons-class strategic laser for ship defense. Several enemy threats are forecast to appear in the next few decades that would overwhelm traditional, kinetic energy–based defenses such as the Phalanx weapons system. For example, a supersonic cruise missile flying low to the water would remain undetected as it approached, cruising beneath radar coverage until popping up in altitude a few miles from a naval ship to position itself for a kill.

With a cruise missile attacking at close distance at supersonic speeds, a ship would not have time to launch its defenses. Even if the ship anticipated the "pop-up" maneuver, fired just as the cruise missile appeared, and somehow managed to hit the missile, the debris would retain its forward momentum and impact the ship.

As a result, in a study in the late 1990s, the Defense Science Board recommended that the navy should consider using an FEL for fleet defense. Although other lasers have demonstrated weapons-class power levels, the FEL was chosen over traditional chemical lasers because they lase at wavelengths easily absorbed by water vapor, which exists in copious amounts near the ocean. Because the FEL's wavelength depends only on the FEL beam energy and wiggler, it can be tuned to lase at an optimum wavelength to minimize absorption by the water vapor.

The Jefferson Lab has since lased at over 10 kilowatts of power, a world record. It is conducting good science with its CW beam and is making impressive advances in FEL design. But is this the best path to a weapon that can be fielded on a ship?

The Jefferson FEL is large. It is several tens of meters long, and much work is needed to scale the laser to weapons-class, or megawatt power levels. A shipboard FEL would take up much space and has other problems as well.

The Jefferson FEL uses cryogenics for superconducting magnets and has very small tolerances for its laser oscillator that may or may not be suitable for volatile shipboard conditions. This technology is not as mature as the COIL laser, for example, that is being installed on the airborne laser, and other approaches have been proposed for building a weapons-class FEL, such as using an extremely high-gain amplifier instead of an oscillator.[13]

The navy is decades away from fielding an FEL on a ship, and a substantial increase in funding is needed to construct a smaller-scale FEL on the order of hundreds of kilowatts to demonstrate the utility and tactics of making such a large investment.

The FEL offers great flexibility in allowing users to optimally tune its wavelength and can give the navy a strategic advantage in defending the

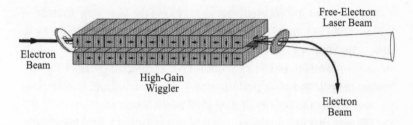

Figure 14.3  Schematic diagram of a free electron laser (FEL). An electron beam is injected into a high-gain (efficient amplifying) wiggler (a series of alternating magnets). Inside the wiggler the electron beam bends or wiggles back and forth, undergoing acceleration and emitting coherent laser radiation. This diagram shows a "one-pass" FEL amplifier cavity without mirrors. (Diagram courtesy Los Alamos National Laboratory, © 2005, the Regents of the University of California)

fleet. However, much research remains to be done, and the navy must make a substantial commitment to this technology if it is to reach the maturity of other directed energy weapons such as the ABL, THEL, and ADS.

A ground-based FEL is much easier to build and would provide a means to use the high-intensity FEL beam to research weapons effects. In fact, the navy would like to use its facility at Barking Sands, Kauai, Hawaii, as a FEL test bed to conduct laser demonstrations in a maritime environment.

Ground-based FELs could use space-based relay mirrors to transmit high-power laser beams anywhere on earth and provide a means to kill ballistic missiles after they have left their boost phase, supplementing ABL's mission. Building the FELs at various locations would ensure that at least one of them would not be stopped because of bad weather.

The first chief scientist of SDI, Dr. Gerry Yonas, now a vice president at Sandia National Laboratory, maintains that this argument for using an FEL for strategic defense was so compelling that the initial construction of an FEL at White Sands Missile Range convinced the Soviet Union that the United States would succeed—and this indirectly contributed to the eventual fall of the USSR and helped win the Cold War.[14]

Although unsubstantiated, this claim shows the sweeping promise of building an FEL. There's something magical about having the ability to tune a high-energy laser to a specific wavelength, and some scientists think the killer app for FELs are not for use as directed energy weapons, but for medical and biological imaging—from peering into human bodies to watching the activity of biological molecules in real time.[15]

Whatever the eventual use, the lure of exploiting a tunable electric laser, one free of noxious chemicals, gas discharges, uncooperative crystals, or messy liquids, is appealing and may eventually be where the future of lasers leads us.

# 15

# The Next Generation:
# Fibers, Phasing, and Terahertz

## Are There Other Types of Lasers?

SO FAR WE'VE SEEN THAT it's possible to create a weapons-class laser—one that can generate megawatts of power—but it's certainly not easy.

In addition to the complex interactions of the various laser systems—the optical train, beam control and stability, target acquisition and tracking, and the laser source itself—each laser comes with its own peculiar set of problems.

The ABL COIL chemical laser carries the baggage of having to transport huge quantities of noxious fuels such as chlorine gas and an aqueous mixture of hydrogen peroxide and potassium hydroxide. Although the military has a long history of coping with hazardous materials under a generally unblemished safety record (nuclear warheads come to mind), they cope because they have to in order to accomplish their mission, not because they want to. Ask any logistics commander. Given a choice between fielding two weapons systems that produce identical results, would you rather deal with a system that had life-threatening chemicals or one that did not? That's a no-brainer. The army's MTHEL carries much the same baggage and risk. The MTHEL's lasing medium is created when fluorine atoms are supersonically mixed with helium and deuterium to generate deuterium fluoride (DF) in an excited state. Transporting and storing noxious chemicals on the battlefield for a limited engagement is dangerous and involves a high element of risk. It involves a sophisticated logistics tail that may extend from the battlefield to the United States and may involve hundreds of personnel and thousands of hours to transport and store the chemicals close to the battlefield.

Just as the air force deals with the risk involved in using ABL laser fuel, the army will do so as well. The army will use MTHEL because it accomplishes a specific mission. And if presented with two systems which accomplish that mission in an identical fashion, but one without the risk and a complicated logistics tail, the army would certainly choose the lower-risk solution.

So what *is* the solution? At lower, tactical powers on the order of 100 kilowatts, the army is funding research in solid-state lasers. Researchers at the Lawrence Livermore National Laboratory have demonstrated a 30-kilowatt solid-state heat capacity laser (SSHCL) for the army and plan to scale those tests to the 100-kilowatt range. ZEUS, the army's mine-neutralizing laser system discussed Chapter 13, uses Livermore's SSHCL to accomplish their mission.

The major drawback to this technology is extracting the heat from the laser system and allowing the laser medium, solid-state crystals, to cool sufficiently and quickly so that the quality of the beam is not affected. Livermore is working on solving these problems, and the army is enthusiastically planning to deploy a higher-power SSHCL laser weapon to the battlefield.

As discussed in Chapter 14, the navy has taken another route in solving its laser-related problem by using an electric laser—the free electron laser—that avoids a logistics tail and the risk associated with chemical fuels. Granted that chemical lasers don't produce the wavelengths necessary to lase through a water-rich atmosphere and a weapons-class FEL is still years away, but the navy is looking at two major types of FELs, a continuous wave (CW) oscillator by the Jefferson Laboratory (demonstrated at 10,000 watts) and a pulsed amplifier designed by the Los Alamos National Laboratory.[1]

Although chemical and electric lasers have different problems, they have one problem in common—these systems are large, some larger than others. If deployed by the navy, the Jefferson Lab's megawatt-class FEL would stretch at least 30 meters (over 100 feet) and would only fit on a ship the size of an aircraft carrier. The ABL COIL can only be airlifted by a 747. And even the army's MTHEL needs to be transported by several trailers to the battlefield.

It may be possible to produce weapons-class laser power, but today's systems require a huge transportation infrastructure to move the weapon, not to mention the associated logistics tail with the fuel.

So with this in mind, wouldn't it be nice to find a lightweight, agile system that creates immense amounts of power but does not take up much space?

As long as we're fantasizing, wouldn't it be nice if it didn't matter how we treated this system? For example, standard optical beam trains must be properly aligned, and they can't be jiggled or the laser will get out of alignment. Wouldn't it be nice if we could just smash it all together or coil it up like a hose and carry the laser system with us, without regard to alignment?

Well, don't get too excited, but there is a possibility that something like this could become available. It's based on laser fiber technology— the same stuff the communication industry has been using for years—and it might be the next big thing in laser weapons.

## What Are Fiber Lasers?

Simply stated, fiber lasers are fiber optical cables with a laser gain medium embedded in the middle. That definition didn't help me much the first time I heard it, so I'll carve it into easy-to-swallow bites.

## What's a Fiber Optic Cable?

A fiber optic cable is similar to copper wire in that they both carry information, but there are fundamental differences between the two. Fiber optic cables carry information in the form of light, while traditional copper wires (or cables) carry that information as pulses of electricity. Both fiber optics and copper wires are protected by an outer flexible, waterproof material that keeps the relatively delicate inside safe from the elements; but that is where the similarities end. Fiber optic cables are much more efficient than copper wires and can carry much more information.

Laser light is injected in one end of the fiber by focusing pulses from a diode laser called an ILD for an injection laser diode. The laser light bounces down the length of the fiber in a phenomenon known as total internal reflection.

Think of a fiber as a long, flexible tube lined with mirrors, like a hall of mirrors in an amusement park, but with the mirrors covering every inch of space of the walls, ceiling, and floor. Now instead of a large hall, imagine a long tube only 1/100,000 of a meter, or 10 microns, in diameter.

Light injected into one end of the fiber propagates down the tube, bouncing at an angle back and forth through the center. Figure 15.1 illustrates an ILD injecting laser light into a fiber optic cable. For simplicity, in this figure the refracting optic that focuses the beam into the fiber is assumed to be part of the injection laser.

The laser beam bounces down the fiber, and in a perfect world, if the inner surface of the fiber were coated with a perfectly reflective material, the beam would propagate forever.

But as with any mirror, some loss occurs because of absorption into the material. Even with a highly reflective surface lining the inside of a fiber, this loss mechanism could quench the laser and cause the beam to die out. To make matters worse, coating the interior of a fiber optic with a highly reflective material is not possible. The challenge of making a highly flexible fiber with a highly reflective internal coating is a difficult materials science challenge—if it is possible at all.

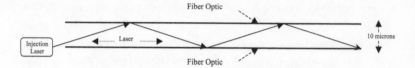

Figure 15.1   A laser beam injected by an injection laser diode bounces down the length of a laser fiber, reflecting off the fiber's inner surface. The beam's injection angle exceeds the critical angle to produce "total internal reflection," allowing the beam to bounce down the fiber with negligible absorption.

So, is a fiber laser another example of something that is theoretically easy to do but results in an impossible engineering task—like Bob Forward's volume of antigravity that according to physics could be demonstrated on earth? Tragically, a thin, flexible means of transmitting laser beams seems just out of reach.

Luckily, a phenomenon known since the advent of early optics—total internal reflection—allows an elegant, easy solution to this apparent stumbling block.

If the laser is injected into the fiber at just the right angle, known as the critical angle that results in total internal reflection, then the beam is almost perfectly reflected and can bounce merrily down a fiber almost forever.[2]

Total internal reflection occurs when the angle that the light grazes inside the surface of the fiber exceeds the critical angle. That angle can be calculated from Fresnel equations, which are derived from Maxwell's equations.

In 1970 Corning Glass Works demonstrated a fiber optic that carried light one kilometer with only a 1 percent loss of energy, the critical limit necessary for fiber to be used commercially. Today, with advances in material science, loss is as low as 0.005 percent per kilometer.

But total internal reflection is still not the silver bullet that ensures a "loss-less" transmission of a laser beam—crucial if fiber optics are to be used as DE weapons.

For example, we've seen that a strategic weapons-class laser system needs to transmit a megawatt of power. If a system based on fiber optics loses even a fraction of that energy, say a thousandth of a percent (0.001 percent), that means thousands of watts of laser power will not be transmitted. That doesn't sound like much, but eventually that extra power, called waste energy, would destroy the transmission quality of the best single optical fiber.

So what do we do? Fiber optics seems ideal. Unlike any other fixed laser optical train that must be immaculately tuned for spatial sensitivity, fibers can be coiled, twisted, and crammed into just about any configuration.[3]

One approach to this problem is similar to a device used by the communications industry applied to traditional copper wire lines—amplifiers (called repeaters) that are inserted periodically throughout the cable to boost the power. When the laser encounters an amplifier, it is magnified in power and can continue down the cable until it encounters another amplifier. This solution works very well for cables that cross intercontinental distances.

Although using repeaters or amplifiers does not solve the problem of waste power destroying the fiber's transmission quality, a variation of this solution embeds a laser gain medium in the fiber itself.

## What's a Laser Gain Medium?

A laser gain medium is anything that amplifies or magnifies a laser.

In a chemical laser, it is the excited molecule that emits a photon in the same frequency, direction, and phase of the laser. In a solid-state laser, such as a crystal, it's the crystal itself. And in an FEL, it's the volume inside the undulating magnets that cause the incident beam to wiggle back and forth. In other words, a gain medium is anything that amplifies the beam.

In a fiber laser, the laser beam passes through the center of the fiber as it propagates down the fiber, bouncing back and forth off the walls. The gain medium is shown in Figure 15.2, inserted in the center of the fiber.

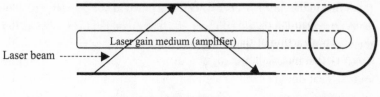

Side view of fiber cable                    Axial view

Figure 15.2    Laser beam bouncing through a center gain (amplifying) medium. Each time the beam bounces through the center, it gains energy and is amplified in power.

Each time the beam bounces off the wall, it passes through this amplifying medium, gaining energy with every bounce. And if this increase in energy is greater than the dispersive losses, then the beam's energy continues to grow.

This bouncing back and forth through the medium is identical to the way a laser oscillates in a traditional laser cavity, bouncing back and forth between two mirrors. The difference is that since the beam is bouncing down the fiber, moving one way down the length, it never goes back and interacts with the same laser medium in the middle. Thus a fiber laser is also known as a fiber amplifier, since the beam is continuously amplified as it bounces down the fiber.

The beauty of a fiber oscillator is that if you want more energy in the beam, all you have to do is to make the amplifier section longer.

But there is a problem. When laser light is amplified in these fibers, so-called nonlinear processes start to occur, which tends to quench or stop lasing. However, these problems are being addressed one by one, and much progress has been accomplished in the past few years, all driven by advances in the telecommunications industry, to generate high-quality beams for transmitting vast amounts of information.

In early 2005, individual fibers not more than a few centimeters in length and about the diameter of a human hair produced 2 kilowatts (2,000 watts) of coherent laser power. Fundamental processes in the lasers prevent transmitting much more power than that, but some researchers think that in the future fibers may carry as much as 10,000 watts (10 kilowatts) a strand. This is still far short of what is required for a weapon, but there are other, tactical applications for such power levels.

And there is more to the story.

## Phasing

It's always possible to gang fiber lasers together. Putting two 1-kilowatt lasers together results in 2 kilowatts of power. However, because the laser radiation was produced in separate fibers, subtle differences in the laser's phases exist, and the total 2-kilowatt beam is no longer coherent.

But if researchers are clever, and if they're careful, they may be able to squeeze more power out of many fiber lasers, through a nonlinear phenomenon known as phasing, than might be available by simply ganging them together. Plus, the total beam is coherent, as if it had been produced in one giant fiber.

Lasers can be phased—precisely matching their wavelengths and phases, much like tuning an instrument in perfect pitch with another instrument.

Even better, when phasing occurs with electromagnetic radiation, the power adds not as a simple linear addition of the number of elements but rather as the *square* of the number of elements.

In other words, two phased lasers produce four ($4 = 2 \times 2 = 2^2$) times more power than that produced by two nonphased lasers; three lasers produce nine ($9 = 3 \times 3 = 3^2$) times the power; and four lasers produce a factor of 16 ($16 = 4 \times 4 = 4^2$) times greater power than four separate lasers.

So in principle, through phasing, it appears possible to reach megawatt power levels using this principle of phasing. However, once again, what may be possible in theory breaks down in reality.

Phasing two lasers is hard, although it's been demonstrated that with much effort phasing four lasers together can be done. But going beyond that number proves extremely difficult.

The main difficulty is that lasers "cross-talk" information, or "bleed" noise from one laser to another that affects phase information and even mode development. Although phasing has been a dream of laser physicists for almost as long as lasers have been around, it is generally agreed that phasing lasers to produce high power will not be very successful.

So what's left? We seem tantalizingly close to having the perfect solution for achieving high power with a simple, flexible system, yet the answer seems frustratingly far away.

Well, there might be an answer, although it is not as elegant as phasing.

I mentioned above that ganging fibers together would result in an increase in power. The lasers won't be in phase, and their beams won't be coherent, but their power is still increased. This phenomenon is true of noncoherent radiation such as that generated by flashlights or spotlights.

Obviously a stage illuminated by four spotlights is four times brighter than one. And although it isn't as bright as it would be if the spotlights were phased, four spotlights is certainly brighter than one.

As opposed to the delicate and difficult job of phasing together even two lasers, it's simple to gang many lasers next to each other and align their beams.

In the future, bundling hundreds of 10,000-watt fibers together would result in power levels of a megawatt—enough for a weapons-class laser. It may be a poor man's version of a coherent beam, but it's good enough to work.

With that capability, the Holy Grail of laser weapons may not be found in 747s carrying tons of laser fuel circling at 40,000 feet, in a series of dusty trailers carrying tons of DF fuel positioned in a semicircle on the desert floor, or even taking up part of the deck on an aircraft carrier, with cryogenic liquids sloshing around with the waves.

Wouldn't it be ironic if laser's megawatt Holy Grail turned out to be a garden-hose thick cable consisting of thousands of strands of slender fiber lasers that could be twisted and contorted into any shape, instead of a "big iron" device that could only be carried on a jumbo-jet or an aircraft carrier?

## Now for Something Completely
## Different: Terahertz (THz)—the Forbidden Band

We've seen that lasers operate in a region bounded by wavelengths ranging from $2 \times 10^{-5}$ cm (ultraviolet) to $1.5 \times 10^{-4}$ cm (infrared), which covers the visible and near infrared spectrum. The frequencies associated with this band range from $10^{13}$ to $10^{15}$ hertz.

It is generally agreed that the microwave regime is bounded by wavelengths from a tenth of a centimeter to several thousand centimeters (0.1 to $10^3$ cm), where the frequencies associated with this band range from tens of megahertz to hundreds of kilohertz.

The region between the far infrared and microwave regime is called the terahertz (THz) band, and generally starts around 30 GHz (0.03 THz). Just as "kilo" designates 1,000 (or $10^3$), mega designates 1,000,000 ($10^6$), and giga a thousand million (a billion, which is 1,000,000,000 or $10^9$), the "tera" prefix designates a thousand billion (equivalently, a million million, or $10^{12}$). Terahertz wavelengths run from 30 μm (or 1/1,000,000 of a meter—a thousand times smaller than a millimeter) to about 1,000 μm, which is a millimeter.

The terahertz band is largely unexploited, mostly due to the many absorption resonances that occur in the molecules that make up the atmosphere. As a result, not much effort has been put into developing terahertz sources and technology. If it's hard to propagate and detect, then why put forth the effort? Also, the atmospheric attenuation is not only large but also extremely variable.

Atmospheric absorption is not the only resonance that contributes to the high attenuation of terahertz radiation. Certain rotational absorptions occur in organic molecules, making the terahertz band ripe for exploiting biomedical applications such as medical imaging and security scanning systems.[4] This makes terahertz radiation appealing for such esoteric applications as detecting pathogens (those nasty little bugs that might be used for weapons of mass destruction), imaging and tracking radar, secure communications, and perhaps even as an aspect of directed energy weapons. Since terahertz wavelengths are shorter than microwaves, they don't diffract as much as HPM; and since their wavelengths are longer than visible lasers, they can couple better with electronic systems, much like HPM. THz seems the ideal type of directed energy to exploit; you can transmit it far distances like a laser while packing the damage of HPM.

So is terahertz radiation the savior of HPM, without the downfalls we discussed earlier?

Not quite. There's one small problem.

Remember how THz radiation can't propagate in the atmosphere because of atmospheric absorption? This is a big problem. So big that it could cause using THz in the atmosphere to be a nonstarter.

Unless THz radiation can be used in space where there is no atmospheric absorption and where its attributes really shine, THz applications may never get out of the starting gate because of the simple fact that it just can't propagate.

But before we give up and declare all is lost, remember that these absorption problems are the reason why the THz band was not vigorously researched when radiofrequency (RF) and laser technologies were making huge strides. In fact, a fascinating property of terahertz radiation occurs because of its unique wavelength between RF and light.

Recall that the small size of a laser wavelength enables a laser beam to interact, or couple, with matter at the molecular level; lasers couple with solids to burn off matter, creating a plasma. It does not penetrate far into things such as building walls or the skins of missiles.

On the other hand, recall that RF wavelengths, especially in the form of high-power microwaves, are too long to directly interact with material properties; instead, they couple with larger features such as cables or electronic circuits that are the same size as the microwaves. Radiofrequency waves thus penetrate into things like walls of buildings.

Every wonder why you can hear a radio inside a building, even if there is no window to let in light? It's a matter of how much coupling, or direct interaction, an electromagnetic wave can have with the walls.

Since THz frequencies are in between RF and the near IR, terahertz radiation can penetrate things like cardboard, clothing, paper, wood, thin walls, ceramics, and people. A few years ago a scientific paper hit the street showing how it is possible to see inside a box using terahertz radiation. In this case, the researchers were able to peek inside a cereal box and see raisins.[5] Aside from helping the breakfast industry finally determine how many raisins are really in a box, this is profound application.

Superman's X ray vision might now become a reality and benefit consumer applications, ranging from ensuring quality control to adding security measures at airports, train stations, or any other place where we might think terrorists may lurk.

Even more significant is the possible application to communications. Since THz radiation can only propagate for a short distance, sometimes as short as a few hundred meters in the atmosphere, THz communication gear could be used as a short-range, covert means of establishing a secure communications channel between troops or other nearby combatants that may not want to be overheard.

This could profoundly impact a battlefield where there are thousands of remote-controlled platforms, especially when a platform has strayed far from the "mother ship" (the controlling communications node), and U.S. forces do not want enemy receivers to detect the position of the controller. The RF signals could still be relayed back to the controller via a distributed link, but the precise location of the controller could always be kept secret.

Another application is in space-to-space communications between satellites. Currently there is a strong effort to use laser communications in satellites to increase the amount of information transmitted to the ground. Lasers can dramatically increase communication bandwidth—the amount of information that can be transmitted. By modern-day estimates, communicating with lasers instead of radio waves might give a 20 to 5,000 GHz bandwidth over current RF systems, an increase of nearly a thousand.

Satellites must also communicate among themselves, but doing this with lasers may not be as wise as using lasers for ground-to-space communications.

Terahertz technologies allow another communication option, especially in free space where atmospheric absorption is not a consideration. One terahertz source has demonstrated the capability to carry as much as 10 percent bandwidth, meaning that for a 1 THz signal, the bandwidth would be 100 GHz, in between that of RF and lasers.

Although this increase is not as high as that promised by a laser, the communications infrastructure would be similar to RF and should be much cheaper than that needed by laser communication. This is because a laser communication system requires the highest degree of accuracy and stability to locate, track, and maintain a high bandwidth communications link with a satellite platform; it also requires an entirely new communications infrastructure to support this effort. Satellite-to-satellite communication is much more difficult than ground-to-satellite communication because both platforms are moving.

The current standard satellite communications system uses a radio-frequency infrastructure of antennas, transmitters, receivers, encryption devices, storage units, and other associated technology. Using lasers instead of RF will add a significant amount of radically different infrastructure and therefore higher cost, especially when communicating between satellites. THz communication may provide a sensible middle ground, exploiting an increase in bandwidth while not radically changing the infrastructure.

Terahertz radiation promises a wide number of applications that we can only dream of today, from medical and biological imaging, to vastly increased bandwidths of communication; from extending the range of radar to allowing covert communications; and from enabling the detection of proliferates associated with weapons of mass destruction, to vastly smaller and more efficient sources for the military's Active Denial System.

And even more exciting is that once the terahertz bandwidth has been cracked and stable sources have been developed, just as any other technological advance, the most far-reaching applications are yet to be discovered.

## Femtosecond Lasers

A discussion of advanced directed energy concepts would not be complete without mentioning femtosecond lasers. This is a relatively new, exciting field that has great promise for directed energy applications.

Unlike terahertz radiation, which is confined to a certain part of the electromagnetic spectrum (typically 300 GHz to 1,000 THz), femtosecond lasers can have any frequency. Femtosecond lasers are defined not by the length of their wave but by the shortness of their pulse.

What are laser pulses? As it turns out, lasers may be classed as continuous or pulsed. CW, or continuous wave lasers, are always "on"—their beams consist of a continuous radiation field. COIL is an example of a laser that has a continuous beam.

Pulsed beams consist of one or more pulses of laser radiation. Solid-state lasers, such as ruby lasers and FEL, are typically pulsed. Sometimes the pulses are so close together that we need to rely on sophisticated diagnostics to tell us if the beam is CW or pulsed.

It's like a movie film. To the audience the film looks continuous, as if the picture flows seamlessly across the screen. But in reality, we know that the film is projected at more than 30 frames a second. Our eyes cannot detect movement faster than that, and to us, the motion picture frames appear smooth. In the same way, if spaced close together in time, laser pulses appear continuous.

Now image a single laser pulse. If that one pulse is shorter than a femtosecond, then the laser that generated the pulse is called a femtosecond laser. Simple.

Okay, but what's a femtosecond?

A femtosecond is a millionth of a billionth of a second, or $10^{-15}$ seconds. That's quick.

To put this in perspective, remember that the atmosphere moves on timescales of a thousandth of a second, or a millisecond ($10^{-3}$ seconds). A thousand times quicker than that, in a microsecond ($10^{-6}$ seconds), light travels a third of a kilometer. And in a nanosecond, a billionth of a second ($10^{-9}$ seconds), light travels a third of a meter—about a foot.

So in a femtosecond, light travels 0.3 millionths of a meter, or 0.3 microns.

Pulses this short and shorter allow researchers to study fundamental phenomena such as inner shell dynamics in atoms, as well as chemical and biological processes. The laser pulses occur so quickly that these short beams can penetrate materials faster than the material can react (technically, scientists say this is faster than the "phonon timescale"), and as a result the material doesn't heat up and is not damaged.

Myriad other phenomena occur because femtosecond laser pulses are so short that they, unlike normal, longer pulse or CW lasers, don't react with the environment. For example, femtosecond lasers create filaments, or stands of plasma channels, when they propagate through the atmosphere. This channeling creates a "self-focusing" effect, which in turn causes the femtosecond pulse not to spread out or diffract like a normal laser, and instead focuses into a much smaller spot than expected. In addition, when femtosecond lasers hit certain metals, they cause a "splash" of radiation, such as terahertz radiation.

So in addition to their obvious diagnostic capability, femtosecond lasers will be used in ways we can't even begin to imagine, including laboratory shortwave (X ray, UV) sources, ultrafast spectroscopy, precision drilling, ultrahigh bandwidth communication, particle acceleration, and perhaps even DE weapons.

These effects have been observed by many researchers, making femtosecond lasers a hot research topic for future directed energy concepts.

# 16

# This Is the End

## The Case for Directed Energy

ADVANCES IN DIRECTED ENERGY SCIENCE and technology, as well as investments in enabling infrastructure such as beam control, have matured DE to the point where it can transition from the laboratory to the battlefield.

Laser power has increased over a billion times in the past four decades, allowing laser weapons to soon be fielded in the air (ABL, ATL), on the ground (MTHEL, ZEUS), and within a decade or so on the sea (FEL).

Research in millimeter wave technology, an extremely short-wave version of the radio frequency spectrum, produced ADS (Active Denial system), the world's first long-range, nonlethal antipersonnel weapon that not only gives war fighters the option for assessing the intent of attackers but also provides a clear option between two wildly disparate options of shouting at people and shooting them.

Advances in high-power microwave science and technology promise weapons in the next few decades that can destroy machines and not people. And as the father of HPM research states, "The smarter the opponent's weapons, the dumber we can make them."[1]

## What Are Directed Energy Weapons?

Directed energy weapons exploit part of the electromagnetic spectrum to create a militarily significant effect—preventing an enemy from conducting operations, either by destroying a target or by stopping the enemy.

In the 1980s, particle beams—electron, ion, or even so-called neutral beams such as hydrogen atoms—were considered part of the directed energy arsenal. One distinction between particle beams and today's definition of DE is that particle beams have nonrelativistic mass and could thus use kinetic energy to help destroy a target.[2] Research in the 1980s and 1990s showed that although particle beams had much destructive power, they were difficult to generate and propagate through air. In a word, they were unstable. And although that stability problem was solved for certain conditions, the power supply needed to produce stable beams made particle beam weapons impractical.

Today DE is generally regarded as lasers and microwaves. In contrast, traditional energy weapons such as bullets and bombs use kinetic energy, or energy of motion, as part of their destructive power, along with the damage caused by high explosives or fragments.

On the other hand, directed energy weapons are bundles of photons, electromagnetic energy at a specific frequency and wavelength. As opposed to the electromagnetic energy emitted by lightbulbs, a flashlamp, or even the sun, DE photons are coherent—they all have the same wavelength, the same phase, and they all travel in exactly the same direction.

## The Difference Between
## Lasers and High-Power Microwave Weapons

Lasers and high-power microwaves are simply different manifestations of the electromagnetic spectrum. They both consist of photons—bundles of electromagnetic energy—and the only difference between the two is that they have different energy, a function of wavelength, or equivalently frequency.

Because of this difference in energy, lasers and high-power microwaves have to be created in different ways. In addition, they propagate through the atmosphere and in space differently, and they interact with targets differently.

Laser wavelengths are as much as 10,000 times smaller than microwaves. As such, they diffract up to 10,000 times less than microwaves, which allows lasers to propagate up to 10,000 times farther than a similarly generated microwave to deposit the same energy on a target. This allows lasers to use high-precision reflecting mirrors to redirect their devastation throughout the battlefield, and with the use of space-based relay mirrors, perhaps even around the globe.

However, lasers scatter in the atmosphere more than microwaves do because the laser wavelength is about the same size as atmospheric gas molecules.

Finally, when a laser hits a target, it tends to heat up the target material and burn away layer after layer, producing a copious plume of plasma. The ABL uses this heat-producing mechanism to weaken the skin of a ballistic missile around the missile's fuel tank, allowing the tank's internal pressure to explode; MTHEL uses its laser's heating mechanism to cause a rocket warhead to explode.

On the other hand, because the wavelength of a high-power microwave is so much larger than that of a laser, the HPM interacts with the electronics and circuits of a weapon. Like a thief slipping into a house through the unlocked backdoor, HPM slithers into a target and wreaks havoc with the electronics, the very stuff that makes modern weapons so lethal.

However, microwaves are not yet ready for prime time. Much work needs to be accomplished increasing microwave power and shrinking the infrastructure that generates HPM; this may take decades, but it is necessary before having a battle-ready HPM weapon.

Active Denial uses a high-frequency version of HPM called millimeter waves to heat up a minuscule depth of a person's skin to create a "flee" effect. It provides an ultraprecise, unique, nonlethal means of countering personnel actions.

Despite the outward dissimilarities between lasers and HPM, their similarities far outweigh their perceived differences:

- They both exploit parts of the electromagnetic spectrum.
- They are both impervious to the effects of gravity or ballistic motion.
- They are both ultraprecise, allowing for enormous amounts of energy to be applied exactly where the war fighter wants. This is in contrast to kinetic energy precision weapons, which although relatively accurate (remember those pictures of 500-pound bombs flying through the window of a military installation in Iraq?) may have devastating, unintended collateral effects—such as death— due to blast and fragments.

And the most important aspect of directed energy weapons is the best feature of all: their speed. Kinetic energy weapons reach their target at the speed of sound, or in the case of ballistic missiles, at velocities up to Mach 20, or 20 times the speed of sound, enabling them to hit targets around the globe on the order of 45 minutes.

But even with perfect knowledge of an enemy's position, in 45 minutes a mobile target can move 100 miles away from its initial position.

In fact, the "actionable" time loop required of fighter jets called in to support ground troops in the second Gulf War was less than 10 minutes—the time a typical mobile missile launcher took to pack up and motor away.

In today's environment of stealthy terrorist camps and fast-moving weapon caravans, that time is much less than a minute.

As such, having an ultraprecise weapon capable of striking around the globe almost instantaneously, in less than a second, provides the technological advantage needed to defeat the next generation of adversaries.

And *that* advantage is only provided by directed energy weapons capable of engaging the enemy at the speed of light.

# Types of Lasers

| Lasing Medium | Type of Laser | Wavelength $\lambda$ nm = nanometer ($10^{-9}$ m) |
|---|---|---|
| FEL | Electric | 0.5 – 300,000 nm |
| $CO_2$ | Gas | 10,600 nm |
| CO | Gas | 5,300 nm |
| DF | Gas | 3,800 nm |
| HF | Gas | 2,800 nm |
| Er: Glass | Solid-state | 1,540 nm |
| COIL (Chemical oxygen-iodine) | Liquid-gas | 1,314 nm |
| Cr: Forsterite | Solid-state | 1,150–1,350 nm |
| HeNe | Gas | 1,152 nm |
| Argon | Gas-ion | 1,090 nm |
| Nd: YAP | Solid-state | 1,080 nm |
| Nd: YAG | Solid-state | 1,064 nm |
| Nd: Glass | Solid-state | 1,060 nm |

| Lasing Medium | Type of Laser | Wavelength $\lambda$ |
|---|---|---|
| | | nm = nanometer ($10^{-9}$ m) |
| Nd: YLF | Solid-state | 1,053 nm |
| Nd: YLF | Solid-state | 1,047 nm |
| InGaAs | Semiconductor | 980 nm |
| Krypton | Gas-ion | 799.3 nm |
| Cr: LiSAF | Solid-state | 780–1060 nm |
| GaAs/GaAlAs | Semiconductor | 780–905 nm |
| Krypton | Gas-ion | 752.5 nm |
| Ti: Sapphire | Solid-state | 700–1000 nm |
| Ruby | Solid-state | 694 nm |
| Krypton | Gas-ion | 676.4 nm |
| Krypton | Gas-ion | 647.1 nm |
| InGaAlP | Semiconductor | 635–660 nm |
| HeNe | Gas | 633 nm |
| Ruby | Solid-state | 628 nm |
| HeNe | Gas | 612 nm |
| HeNe | Gas | 594 nm |
| Cu | Metal vapor | 578 nm |
| Krypton | Gas-ion | 568.2 nm |
| HeNe | Gas | 543 nm |
| DPSS | Semiconductor | 532 nm |
| Krypton | Gas-ion | 530.9 nm |
| Argon | Gas-ion | 514.5 nm |
| Cu | Metal vapor | 511 nm |
| Argon | Gas-ion | 501.7 nm |
| Argon | Gas-ion | 496.5 nm |
| Argon | Gas-ion | 488.0 nm |
| Argon | Gas-ion | 476.5 nm |
| Argon | Gas-ion | 457.9 nm |
| HeCd | Gas-ion | 442 nm |
| N2+ | Gas | 428 nm |

| Lasing Medium | Type of Laser | Wavelength λ |
| --- | --- | --- |
|  |  | nm = nanometer $(10^{-9} \text{ m})$ |
| Krypton | Gas-ion | 416 nm |
| Argon | Gas-ion | 364 nm  UV (A) |
| XeF | Gas (excimer) | 351 nm  UV (A) |
| N2 | Gas | 337 nm  UV (A) |
| XeCl | Gas (excimer) | 308 nm  UV (B) |
| Krypton SHG | Gas-ion/BBO crystal | 284 nm  UV (B) |
| Argon SHG | Gas-ion/BBO crystal | 264 nm  UV (C) |
| Argon SHG | Gas-ion/BBO crystal | 257 nm  UV (C) |
| Argon SHG | Gas-ion/BBO crystal | 250 nm  UV (C) |
| Argon SHG | Gas-ion/BBO crystal | 248 nm  UV (C) |
| KrF | Gas (excimer) | 248 nm  UV (C) |
| Argon SHG | Gas-ion/BBO crystal | 244 nm  UV (C) |
| Argon SHG | Gas-ion/BBO crystal | 238 nm  UV (C) |
| Argon SHG | Gas-ion/BBO crystal | 229 nm  UV (C) |
| KrCl | Gas (excimer) | 222 nm  UV (C) |
| ArF | Gas (excimer) | 193 nm UV (C) |

# Notes

## Directed Energy Milestones

1. Shigen Nakajima, "Japanese Radar Development Prior to 1945," *IEEE Antennas and Propagation,* December 1992. This article describes a Japanese-designed magnetron for use as a high-power microwave weapon. The magnetron emitted 100 kilowatts (continuous wave) microwaves at a wavelength of 20 centimeters.

## Chapter 1

1. Actual accuracy numbers are inferred from the *United States Strategic Bombing Survey: Summary Report (Pacific War), July 1, 1946* (Washington, D.C.: United States Government Printing Office, 1946). The Strategic Bombing Survey gives gross numbers that range from 10 percent of bombs hitting the target area (250 to 1,000 feet from target) to 50 percent for low-altitude, carrier-based planes.

2. *Biography of Astronaut General Thomas P. Stafford,* National Aeronautics and Space Administration, Lyndon B. Johnson Space Center, Houston, Texas 77058.

## Chapter 2

1. Al Kehs, "Introduction to RF-DEWs," Directed Energy Professional Society short course, Rockville, Md., October 2004, p. 6.

2. Actually, referring to light in terms of wavelength or frequency is rather simple. Some physicists and chemists refer to light in terms of "wave number," the number of waves that can fit in a certain period; some refer to regions of the electromagnetic spectrum, such as the "W" band (terahertz). A short taxonomy of these bands is given later.

## Chapter 3

1. This quote is taken from Gregory H. Canavan, Nicholaas Bloembergen, and C. Kumar Patel, "Debate on APS Directed-Energy Weapons Study," *Physics Today*, November 1987, p. 48. In this article Canavan refutes the APS study and is debated by Bloembergen and Patel. Also note that nearly 20 years of research has been conducted in directed energy since the publication of this article, and major, strategic DE systems are just now being fielded.

2. Part of the following discussion is taken from J. D. Beason, *DoD Science and Technology: Strategy for the Post–Cold War Era* (Washington, D.C.: National Defense University Press, 1997), pp. 12–15.

3. Kenneth L. Adelman and Norman R. Augustine, *The Defense Revolution: Intelligent Downsizing of America's Military* (San Francisco: Institute for Contemporary Studies Press, 1990), p. 55; and CNN for the second Gulf War.

4. J. F. C. Fuller, *Armament and History* (New York: Scribners, 1945), p. 7.

5. Simon P. Worden, *SDI and the Alternatives* (Washington, D.C.: National Defense University Press, 1991), pp. 13–15.

6. Worden, *SDI*, p. 14.

## Chapter 4

1. Which the author's daughters assure him he is.

2. This problem of throwing S&T over the transom is discussed in detail in Beason, *DoD Science and Technology: Strategy for the Post–Cold War Era* (Washington, D.C.: National Defense University Press, 1997). The easiest solution is to require the innovator to remain as the S&T champion throughout the life cycle of taking the invention to the battlefield, thus ensuring that the problems encountered in fielding the weapon are fixed by the scientist or engineer responsible for inventing the technology. But this solution is hardly ever followed, as process "stove-pipes" of S&T, R&D, testing, and manufacturing are decoupled from interacting.

3. Kenneth L. Adelman and Norman R. Augustine, *The Defense Revolution: Intelligent Downsizing of America's Military* (San Francisco: Institute for Contemporary Studies Press, 1990), p. 66.

4. Adelman and Augustine, *Defense Revolution*, p. 66.

5. Dr. Kirk Hackett, conversation with author, Air Force Research Laboratory, January 15, 2005.

6. Hackett, conversation with author, January 15, 2005.

7. Dr. Bill Baker, chief scientist of the Air Force Research Laboratory's Directed Energy Directorate.

8. Best-selling author Kevin J. Anderson and Doug Beason had a proposed novel based on an HPM attack on Wall Street rejected after just having sold a novel to Universal Studios—meaning that in the early 1990s an HPM weapon was considered "too science fiction" even by Hollywood standards, which illustrates that truth is indeed stranger than fiction.

## Chapter 5

1. "Near real-time" response does not mean an instant response. Instead, it is defined by the U.S. military as being within the "cycle time" of the enemy—the time it takes to produce a critical function. In the first Iraq War, it took a minimum of 10 minutes to erect a Scud launcher against U.S. forces. If the Iraqis erecting the missile could be detected and destroyed within this 10-minute time frame, this is good enough to be called in near real-time.

## Chapter 6

1. H. Motz et. al., *Journal of Applied Physics* 24 (1953): 826.

2. U.S. Patent no. 2,929,922.

3. We shall see in Chapter 7 that being "merely an engineering problem" sometimes means being impossible to accomplish.

4. Gregory H. Canavan and John F. Lilley, "Managing Public Sector Research and Development: Innovation Versus Responsiveness," in *Defense Technology*, ed. A. Clark IV and J. Lilley (New York: Praeger, 1989), p. 244.

5. T. H. Maiman, "Stimulated Optical Radiation in Ruby," *Nature*, August 6, 1960, pp. 493–494; John A. Osmundsen, "Light Amplification Claimed by Scientist," *New York Times*, July 8, 1960, pp. 1, 7.

6. Arthur L. Schawlow, "Lasers: The Practical and the Possible," *Stanford Magazine*, Spring-Summer 1979, pp. 24–29.

7. High-Energy Laser Short Course, John Albertine, consultant.

8. *US Corporate R&D Investment, 1994–2000 Final Estimates*, Office of Technology Policy, Bureau of Economic Analysis, U.S. Department of Commerce, March 28, 2002.

9. Joe Miller, communication with author, January 17, 2005.

10. An oscillator is a volume in which the laser oscillates back and forth, bouncing off mirrors on either end. The laser builds in intensity with each bounce through the

oscillator. An amplifier is typically a "one-pass oscillator." The laser gains so much intensity as it passes through the cavity that it only needs to traverse the amplifier once before it is powerful enough to leave the cavity.

11. H. W. Babcock, "The Possibility of Compensating Astronomical Seeing," *Publ. Astron. Soc. Pac.* 65 (1953): 229–236.

12. James A. Abrahamson and Henry F. Cooper, *What Did We Get for Our $30-Billion Investment in SDI/BMD?* National Institute for Public Policy, 1999.

13. C. Patel and N. Kumar, "Selective Excitation Through Vibrational Energy Transfer and Optical Laser Action in $N_2$-$CO_2$," *Physical Review Letters* 13 (1964): 617–619.

14. N. G. Basov and A. N. Oraevskii, "Soviet Physics," *JETP* 17 (1963): 1171.

15. E. T. Gerry, *IEEE Spectrum* 7 (1970): 51.

16. Dr. Michael D. Griffin, general manager, Advanced Systems Group and executive vice president, Orbital Sciences Corporation, in remarks before the National Security Industry Study Symposium Space: Getting Out of the Box, Arlington, Va., March 6, 1996.

17. ARPA—DARPA's predecessor—was established on February 7, 1958, as a direct response to the Soviet launch of *Sputnik* on October 4, 1957.

18. On February 26, 1962, the Air Force Special Weapons Center located at Kirtland AFB, New Mexico, was awarded $800,000 from ARPA to investigate the feasibility of constructing a high-energy laser.

19. Dr. Joe Miller was one of several laser pioneers in industry in the 1960s and 1970s, including Ted Jacobs, Willy Behrens, Don Bullock, Dale Hook, John Waypa, Don Winter, John Stansel, Ray DeLong, Alvin Schnurr, and Grant Hosack to name a few. Many other people in the military, government, and academia contributed as well. Since it is not possible for me to highlight every individual who made a contribution to high-energy lasers, I use Dr. Miller as an "everyman" to personalize and highlight the high-energy milestones.

20. Dr. Joe Miller, interview by author.

21. James A. Abrahamson and Henry F. Cooper, *What Did We Get for Our $30-Billion Investment in SDI/BMD?* National Institute for Public Policy, 1999.

## Chapter 7

1. Director, defense research and development briefing, May 2004.

2. $CO_2$ lasers reached megawatt power levels in 1972. Robert W. Duffner, *Airborne Laser: Bullets of Light* (New York: Plenum, 1997). The navy's mid-infrared advanced chemical laser (MIRACL) DF laser reached these levels in 1980 at the White Sands Missile Range.

3. Mark E. Rogers, *Lasers in Space: Technological Options for Enhancing U.S. Military Capabilities*, Occasional Paper no. 2, Center for Strategy and Technology, Air War College, Maxwell Air Force Base, Alabama, November 1997.

4. See, for example, the Weinberger-Powell doctrine.

5. Robert W. Duffner, Robert R. Butts, J. Douglas Beason, and Ronald R. Fogleman, "Directed Energy: The Wave of the Future," in *USAF Centennial Celebration of 100 Years of Flight* (2003).

6. The author had the opportunity to brief the air force chief of staff, General Michael E. Ryan, in the late 1990s, on the support needed to maintain the high-energy laser industrial base, and this question was asked.

7. The situation with commercial industry refusing not to yield to government demands became so bad that the USAF once demanded that a major telecommunications contractor modify its accounting practices so it could accept a U.S. government check for a *profit* on the order of $500,000. The contractor decided that it would not be worth modifying its accounting procedures and refused the check and made it a company policy not to undertake government contracts in the future.

8. George M. Dryden, "The C-17 Transport," *Joint Forces Quarterly*, Summer 1996, pp. 67–73.

9. *U.S. Corporate R&D Investment, 1994–2000 Final Estimates*, Office of Technology Policy, Bureau of Economic Analysis, U.S. Department of Commerce, March 28, 2002.

10. John B. Wissler, "Organization of the Joint Technology Office," *Program Manager*, November-December 2002, pp. 26–31.

## Chapter 8

1. Shigen Nakajima, "Japanese Radar Development Prior to 1945," *IEEE Antennas and Propagation*, December 1992. This article describes a Japanese-designed magnetron for use as a high-power microwave weapon. The magnetron emitted 100 kilowatts (continuous wave) microwaves at a wavelength of 20 cm.

2. See Robert W. Duffner, *Airborne Laser: Bullets of Light* (New York: Plenum, 1997), for an excellent review of the airborne laser and a detailed history of the early air force high-power laser program.

## Chapter 9

1. Dr. Kirk Hackett (former ADT program manager), interview by Dr. Bob Duffner, October 26, 2001.

2. For some examples of the Active Denial-like bioeffect literature, see James R. Jauchem, "A Literature Review of Medical Side Effects from Radio-Frequency Energy in the Human Environment: Involving Cancer, Tumors, and Problems of the Central Nervous, System," *Journal of Microwave Power and Electromagnetic Energy* 38, no. 2 (2003): 103–124; Jauchem, Kathy L. Ryan, and Melvin R. Frei, "Cardiovascular and Thermal Responses in Rates During 94 GHz Irradiation," *Bioelectromagnetics* 20 (1999): 264–267; Steven Chalfin, John A. D'Andrea, Paul D. Comeau, Michael E. Belt, and Donald

Hatcher, "Millimeter Wave Absorption in the Nonhuman Primate Eye at 35 GHz and 94 GHz," *Health Physics* 83, no. 1 (2002): 83–90; Andrei G. Pakhomov, Yahya Akyel, Olga N. Pakhomova, Bruce E. Stuck, and Michael R. Murphy, "Review Article: Current State and Implications of Research on Biological Effects of Millimeter Waves: A Review of the Literature," *Bioelectromagnetics* 19 (1998): 393–413; Thomas J. Walters, Kathy L. Ryan, David A. Nelson, Dennis W. Blick, and Patrick A. Mason, "Effects of Blood Flow on Skin Heating Induced by Millimeter Wave Irradiation in Humans," *Health Physics*, February 2004, pp. 115–120; D. A. Nelson, T. J. Walters, K. L. Ryan, K. B. Emerton, W. D. Hurt, J. M. Ziriax, L. R. Johnson, and P. A. Mason, "Inter-Species Extrapolation of Skin Heating Resulting from Millimeter Wave Irradiation: Modeling and Experimental Results," *Health Physics*, May 2003, pp. 608–615; Patrick A. Mason, Thomas J. Walters, John Di-Giovanni, Charles W. Beason, James R. Jauchem, Edward J. Dick Jr., Kavita Mahajan, Steven J. Dusch, Beath A. Shields, James H. Merritt, Michael R. Murphy, and Kathy L. Ryan, "Lack of Effect of 94 GHz Radio Frequency Radiation Exposure in an Animal Model of Skin Carcinogenesis," *Carcinogenesis* 22, no. 10 (2001): 1701–1708; Kathy L. Ryan, John A. D'Andrea, James R. Jauchem, and Patrick A. Mason, "Radio Frequency Radiation of Millimeter Wave Length: Potential Occupational Safety Issues Relating to Surface Heating," *Health Physics*, February 2000, pp. 170–181; Nelson, D. A., M. T. Nelson, T. J. Walters, and P. A. Mason, "Skin Heating Effects of Millimeter-Wave Irradiation: Thermal Modeling Results," *IEEE Transactions on Microwave Theory and Techniques*, November 2000, pp. 2111–2120; Kenneth R. Foster, John A. D'Andrea, Steven Chalfin, and Donald J. Hatcher, "Thermal Modeling of Millimeter Wave Damage to the Primate Cornea at 35 GHz and 94 GHz," *Health Physics*, June 2003, pp. 764–769; Dennis W. Blick, Eleanor R. Adair, William D. Hurt, Clifford J. Sherry, Thomas J. Walters, and James H. Merritt, "Thresholds of Microwave-Evoked Warmth Sensations in Human Skin," *Bioelectromagnetics* 18 (1997): 403–409.

3. Blick et al., "Thresholds," pp. 403–409.

4. The Directed Energy Bioeffects Division is headquartered at Brooks City Base, San Antonio, Texas. The division is part of the Air Force Research Laboratory's Human Effectiveness Directorate. Its mission is to predict, mitigate, and exploit the bioeffects of directed energy on Defense Department personnel, aerospace missions, and the environment; recommend safety standards and provide system design specifications; conduct health and safety analyses and bioeffects validation for proposed non-lethal technologies, http://www.he.afrl.af.mil/Div/index.htm#HED.

5. Dr. Kirk Hackett, interview by Dr. Bob Duffner, October 26, 2001.

6. Hackett, interview by Duffner, October 26, 2001.

7. Robert S. Dudney, "Hans Mark Looks Ahead," *Air Force Magazine*, January 1999.

8. Robert W. Duffner, *Airborne Laser: Bullets of Light* (New York: Plenum, 1997).

9. Hackett, interview by Duffner, October 26, 2001.

10. Lieutenant Colonel Chuck Beason, Active Denial technology program manager, at Brooks AFB, Texas.

11. Hackett, interview by Duffner, October 26, 2001.

12. The author was present at that dinner, along with Dr. Good, Dr. Hackett, and Dr. Bill Baker, then chief scientist of the DE Directorate.

13. "Marine Corps Times Reports on Secret Weapon Program," *Army Times*, February 23, 2001.

14. "New Technology Drives Away Adversaries," Joint USMC-USAF press release, USMC no. 2001–09, DE release no. 2001–09, February 22, 2001.

15. The Force Protection Battle Lab was charged with coordinating the overall testing, and a series of formal phases with exit criteria were established to judge the veracity of using Active Denial in combat. The Battle Lab was an air force unit headquartered in San Antonio and chartered to conduct quick-reaction tests that could be transitioned to the field. Phase 1 included standard "force-on-force" modeling and simulation that assessed using Active Denial to defend various assets in different scenarios. Phase 2 used humans in similar scenarios, but with a simulated Active Denial beam. The first part of Phase 3 was to measure the desired flux of millimeter waves at the preapproved distance from the source, and to assess human response.

16. Dr. Diana Loree, interview by Dr. Barron K. Oder, March 3, 2004.

17. The events of the first ADS test were related to the author by Lieutenant Colonel Chuck Beason and Dr. Kirk Hackett in several conversations.

18. The test subject identities are protected by Privacy Act restriction. In addition to the VIPs, all scientists and engineers on the project, as well as the senior leadership of AFRL's Directed Energy Directorate underwent the Active Denial testing.

19. The popularly stated mission of the air force is to "Fly and Fight, and Don't You Forget It," not to use nonlethal force to stop, delay, deter, and turn back insurgents. As a result, ADS has had a hard time gaining upper-level USAF support until the air force realized that protecting bases (under force protection) was core to its mission.

20. These views belong to the author, and not to Dr. Loree or anyone else in the USAF.

21. Major General (Dr.) Donald Lamberson, interview by author, November 6, 2004.

22. Lamberson, interview by author, November 6, 2004.

23. Lamberson, interview by author, November 6, 2004.

24. Bernard C. Nalty, ed., *Winged Shield, Winged Sword: A History of the United States Air Force*, vol. 1, *1907–1950* (Washington, D.C.: Government Printing Office, 1997), p. 409.

## Chapter 10

1. Robert W. Duffner, *Airborne Laser: Bullets of Light* (New York: Plenum, 1997).

2. Major General (Dr.) Donald Lamberson, interview by author, November 6, 2004.

3. Before Dr. Teller passed away, it was well-known around scientific circles that Edward (affectionately known as "Yoda" because of his hunched appearance and ubiquitous cane) did not like the term "father of the H-bomb." Thus colleagues took care

when speaking of this living legend. However, one story related Dr. Teller being introduced as the "mother of the H-bomb," creating an incredible stir.

4. Lamberson, interview by author, November 6, 2004.

5. Air-to-air Sidewinder missiles were launched at the ALL during tests over the Pacific Ocean off the coast of California. The ALL successfully shot down all missiles, proving that a laser weapon can act as a defensive shield for "high-value" assets. It should be noted that the USAF did not put the ALL crew in danger. Although the missiles were launched by fighter warplanes, the Sidewinders did not have enough fuel to reach the ALL. If the laser test failed, the missiles would fall harmlessly into the ocean.

6. Norman Friedman, *Desert Victory: The War for Kuwait* (Annapolis, Md.: Naval Institute Press, 1991).

7. The time for an Iraqi Scud launcher to come out from behind camouflage, erect its launcher, and light off a Scud was estimated to be on the order of 10 minutes—the so-called cycle time of an enemy tactical ballistic missile threat.

8. The term "traditional air force" has come to mean whoever controls the most power or influence. From World War II to the 1970s, the traditional air force meant Strategic Air Command, which controlled the bombers and the nuclear weapons. In the 1970s, the air force saw a distinct shift. Fighter pilots wrestled control away from the bomber mafia and became the "traditional air force." Although a fighter pilot by early training, Fogelman was also schooled in transport aircraft and gained an appreciation for less glamorous but just as important flying and support activities.

9. Official USAF programs are tracked using program element (PE) codes. Each PE has a program manager in the field, such as the ABL SPO director, and a program element monitor (PEM) in the Pentagon.

10. Colonel Richard Tebay (USAF, ret.), interview by author, September 11, 2004.

11. The Pentagon's office of PA&E (Program Analysis and Evaluation, an independent oversight office reporting directly to the Secretary of Defense) served as a watchdog and aggressively held the USAF's ABL program to the high standards expected of every weapons system. Although there appeared to be much antagonism between the ABL and PA&E, the dynamic tension ensured that the ABL program office was technically ready to proceed in every step.

12. Tebay, interview by author, September 11, 2004.

13. At the 2004 board of directors meeting of the Directed Energy Professional Society, of which the author was then vice president, most board members—some of them intimately connected with the ABL program—stressed that to ensure success, the S&T base for the ABL needed to be expanded. At that time the ABL program was under intense pressure to reach its milestones and needed all its funding to reach its technical and programmatic objectives; thus, there was no additional funding to be invested in expanding the S&T base.

14. Despite her groundbreaking acquisition reforms in the USAF, known as "lightning bolts," that helped streamline the burdened acquisition process, after Druyan

retired from the civil service she was sentenced to prison for her illegal dealings with a major aerospace contractor while still in the government.

## Chapter 11

1. This scenario is taken from a published (public) Air Force Research Laboratory conjecture of using the ABL for other additional, ancillary purposes such as ISR (intelligence, surveillance, and reconnaissance) or cruise missile defense.

2. According to the coast guard, Loran was developed to provide radio navigation for U.S. coastal waters and was later expanded to include complete coverage of the United States, including most of Alaska. Users can return to previously determined positions with an accuracy of 50 meters or better using Loran-C in the time difference repeatable mode—still not accurate enough to allow a plane to land itself.

3. Part of the following discussion is taken from Beason, *DoD Science and Technology: Strategy for the Post–Cold War Era* (Washington, D.C.: National Defense University Press, 1997), pp. 29–31.

4. Nathan Rosenberg and L. E. Birdzell, *How the West Grew Rich* (New York: Basic, 1986), p. 243.

5. Isaac Asimov, *Asimov's Chronology of Science and Discovery*, (New York: Harper & Row, 1989), pp. 338–371.

6. Donald L. Losman and Shu-Jan Liang, *The Promise of American Industry: An Alternate Assessment of Problems and Prospects* (Westport, Conn.: Quorum, 1990), p. 104.

7. Edward A. Duff and Dr. Donald C. Washburn, "The Magic of Relay Mirrors," presented at the Maui Conference on Adaptive Optics, Kirtland AFB, New Mexico, 2004.

## Chapter 12

1. Air Force Research Laboratory.

2. But not "DARPA tough," a term the Defense Advanced Research Projects Agency—the Pentagon's go-to guys for advanced R&D solutions—uses for projects that have a high payoff for national security but also a 50 percent chance of failure: "no pain, no gain." Also, graduate students in astronomy routinely track objects with this accuracy.

3. Over the horizon radar (OTH) uses the phenomenology of bouncing radar signals off the earth's ionosphere to detect targets that are beyond the "line of sight" of a sensor. Much distortion occurs, but this methodology was (and still is) used for detecting missile threats against the United States. However, this is a massive, ground-based system too large for an aircraft carrier to carry, much less an airplane.

4. In practice, the high-altitude airship's mirrors could not reflect both an MTHEL laser and an ABL laser, as the two lasers are different wavelengths. The MTHEL's HF laser is 2.8 microns while the ABL's COIL is 1.314 microns. Although this difference seems

slight, the reflecting mirrors and beam control system for each weapon must be opti-
mized for the system to ensure the highest reflection and lowest beam losses. Thus, if an
MTHEL and ABL both want to extend their range in the battlefield, either additional
HAAs must be present or multiple, unique mirror systems must be on board an HAA.

## Chapter 13

1. Northrop Grumman, press release, August 21, 2003.

2. Northrop Grumman, press release, May 6, 2004.

3. Northrop Grumman, press release, August 21, 2003.

4. Northrop Grumman, press release, August 21, 2003.

5. John M. Spratt (D-SC) ranking member of the House Budget Committee, senior
member of the House Armed Services Committee, and cochair of the Congressional
Electronic Warfare (EW) working group.

6. Jack Urso, Military Information Service, December 10, 2003.

7. Rich Tuttle, "Large Aircraft Infrared Countermeasures System," *Aerospace Daily
and Defense Report,* 2004.

8. Tracy Bunko, "Air Force Approves Initial Production of Laser-Based Jammer Slated
for C-17, C-130," Air Force Material Command news release no. 0832, August 23, 2002.

9. Tuttle, "Large Aircraft."

10. "Humvee" is how the army pronounces HMMWV (high-mobility multipurpose
wheeled vehicle), a light (1.25 ton), diesel-powered, four-wheel-drive vehicle used for
combat operations.

11. Jim Wilson, Beyond Bullets, http://www.popularmechanics.com.science/military,
accessed October 12, 2004.

12. ZEUS-HLONS, HMMWV Laser Ordinance Neutralization System, United States
Army Space and Missile Defense Command, Space and Missile Technical Center, Public
Affairs Office, no. 0183, February 2004.

13. ZEUS-HLONS, HMMWV Laser Ordinance Neutralization System, United States
Army Space and Missile Defense Command, Space and Missile Technical Center, Public
Affairs Office, no. 0183, February 2004.

## Chapter 14

1. Shigen Nakajima, "Japanese Radar Development Prior to 1945," *IEEE Antennas
and Propagation,* December 1992. This article describes a Japanese-designed magnetron
for use as a high-power microwave weapon. The magnetron emitted 100-kilowatt (continuous wave) microwaves at a wavelength of 20 centimeters.

2. Witness the powerful, ground-based radars fielded in the 1950s by both the United
States and the Soviet Union, first for air defense and later to detect incoming ICBMs.

Some of these radars produce electromagnetic energy in the microwave band, and are used today in applications that range from profiling atmospheric winds to tracking aircraft. Early radars in the 1940s generated hundreds of kilowatts, and their power has increased through the years.

3. Dr. Bill Baker, chief scientist of the Air Force Research Laboratory's Directed Energy Directorate.

4. U. Jordan, V. E. Semenov, D. Anderson, M. Lisak, and T. Olsson, "Microwave Breakdown in Air for Multi-Carrier, Modulated or Stochastically Time Varying RF Fields," *Journal of Physics D: Applied Physics* 36 (2003): 861–867.

5. Radio Wave Bands

| Radio Band | | Frequency | Wavelength |
| --- | --- | --- | --- |
| VLF | Very low frequency | 3–30 KHz | 100 km |
| LF | Low frequency | 30–300 KHz | 10 km |
| MF | Medium frequency | 300 KHz–3 MHz | 1 km |
| HF | High frequency | 3–30 MHz | 100 m |
| VHF | Very high frequency | 30–300 MHz | 10 m |
| UHF | Ultra high frequency | 300 MHz–3 GHz | 1 m |
| SHF | Super high frequency | 3–30 GHz | 1 cm |
| EHF | Extremely high frequency | 30–300 GHz | 0.1 cm |

6. See, for example Albert Einstein, *Relativity: The Special and General Theory*; if you have access to the Internet, type GRAVITATIONAL LENS in any popular search engine.

7. J. M. Madey, *Journal of Applied Physics* 42 (1971): 1906.

8. A "wiggler" is a series of alternating magnets that make a charged electron beam wiggle back and forth. When the beam wiggles, it accelerates in a direction perpendicular to the direction of the beam. When the beam accelerates, it emits radiation, photons that are coherent with other photons present in the wiggler.

9. Patrick G. O'Shea and Henry P. Freund, "Free-Electron Lasers: Status and Applications," *Science*, June 8, 2001, p. 1853.

10. These pioneers of HPM research are also called "Marx-bank" scientists, after the high-energy, pulsed Marx capacitor they typically used. A Marx bank is a high-energy storage capacitor that charges in parallel and discharges in series, allowing for pulsed, extremely high currents, on the order of millions of amperes.

11. H. Motz et al., *Journal of Applied Physics* 24 (1953): 826.

12. R. M. Phillips, *Nuclear Instrument Measurements*, vol. A272, 1, 1088.

13. This is not meant to imply that a COIL or any other chemical laser should be considered for use in a maritime environment. An FEL is definitely superior but is not as mature as existing weapons-class chemical lasers. The point is that all FEL designs should be considered before the navy embarks on a multibillion dollar investment.

14. Dr. Gerry Yonas, private communication with author, October 2003.

15. D. M. Mittleman, M. Gupta, R. Neelamani, R. G. Baraniuk, J. V. Rudd, and M. Koch, "Recent Advances in Terahertz Imaging," *Applied Physics B: Lasers and Optics,* 1999.

## Chapter 15

1. One FEL concept the navy is pursuing uses supercooled fluids to enable supercon-ductivity in the free electron laser. However, choosing this path, which uses superconducting magnets to manipulate fields in ultraprecise oscillators, and employing this system on ships will lead to a problem dealing with liquid hydrogen, which is ex-tremely volatile and difficult to manage. In essence, the navy would not be exploiting the FEL for its main inherent advantage over chemical lasers.

2. Again, not really. There is a fundamental limit as to how far the beam will go, as even total internal reflection results in losses to absorption, scattering, and, worst of all, dispersion (spreading out the beam's frequency), which builds up after many kilometers.

3. Recall that after a laser beam is generated by either an oscillator or an amplifier, the beam must be "conditioned" in some way to take out variations across its radius and along its axis. The typical way of doing this involves setting up massive optical tables weighing many hundreds of pounds so that the tables are stable against any outside vi-brations. Providing a lightweight, stable optical bench for the ABL that could simultaneously condition the beam and prepare it for propagating through the changing atmosphere (through adaptive optic technics) was one of the major problems of the ABL program.

4. "Revealing the Invisible," *Science,* August 16, 2002.

5. D. M. Mittleman, M. Gupta, R. Neelamani, R. G. Baraniuk, J. V. Rudd, and M. Koch, "Recent Advances in Terahertz Imaging," *Applied Physics B: Lasers and Optics,* 1999.

## Chapter 16

1. Dr. William Baker, chief scientist, Air Force Research Laboratory's Directed Energy Directorate, Kirtland AFB, New Mexico.

2. Physicists can calculate a mass for light—photons—that is somewhat different from the popularly accepted definition of mass (or weight) that a person can measure by picking up an object. A photon's mass is extremely small and is equivalent to its energy divided by the speed of light squared (from Einstein's $E = mc^2$).

# Glossary

| | |
|---|---|
| ABL | Airborne laser |
| ADS | Active Denial system |
| ADT | Active Denial technology |
| AF | Air force |
| AFRL | Air Force Research Laboratory |
| ALL | Airborne Laser Laboratory |
| Alpha | DARPA's TRW-built DF laser at San Juan Capistrano, a megawatt-class space-based laser demonstrator |
| AO | Adaptive optics |
| APS | American Physical Society |
| ARL | Army Research Laboratory |
| ARM | Aerospace relay mirror system |
| ARPA | Advanced Research Projects Agency, DARPA's predecessor |
| ASC | Air Systems Command |
| ATCD | Advanced tactical concepts demonstration |
| ATL | Advanced tactical laser |
| AWACS | Airborne warning and control system |
| BCS | Beam control system |
| BDL | Baseline demonstration laser |

| | |
|---|---|
| BHP | Basic hydrogen peroxide |
| BILL | Beacon illuminator laser |
| BMD | Ballistic missile defense |
| CAP | Combat air patrol |
| cm | Centimeter |
| $CO_2$ | Carbon dioxide |
| COIL | Chemical oxygen-iodine laser |
| CPB | Charged particle beam |
| CW | Continuous wave |
| DARPA | Defense Advanced Research Projects Agency |
| DDR&E | Director of defense research and engineering |
| DE | Directed energy |
| DEW | Directed energy weapon |
| DF | Deuterium fluoride laser |
| DM | Deformable mirror |
| DoD | Department of Defense |
| DOE | Department of Energy |
| EAGLE | Evolutionary advanced global laser experiment |
| EDL | Electric discharge laser |
| EM | Electromagnetic |
| EMRLD | Excimer midpower raman-shifted laser device |
| FEL | Free electron laser |
| Femtosecond | $10^{-15}$ seconds, or a million billionths of a second |
| FFT | Fast Fourier transform |
| GDL | Gas dynamic laser |
| GHz | Gigahertz ($10^9$ hertz, or a billion oscillations per second) |
| GMTI | Ground moving target indicator |
| HAA | High-altitude airship |
| HELSTF | High-Energy Laser System Test Facility (White Sands Missile Range) |
| HF | Hydrogen fluoride laser |
| HPM | High-power microwave |
| IED | Improvised explosive device |
| ILD | Injection laser diode |
| IRST | Infrared seek tracker |

| | |
|---|---|
| KEKV | Kinetic energy kill vehicle |
| KHz | Kilohertz ($10^3$ hertz, or a thousand oscillations per second) |
| KKV | Kinetic kill vehicle |
| LM | Laser module |
| MASER | Microwaves amplified by the stimulated emission of radiation, a microwave laser |
| MHz | Megahertz ($10^6$ hertz, or a million oscillations per second) |
| Micron | A millionth ($10^{-6}$) of a meter, also written as μm |
| Microsecond | A millionth ($10^{-6}$) of a second, also written as microsec or μsec |
| MIRACL | Navy's mid-infrared advanced chemical laser, located at HELSTF |
| MSSC | Maui Space Surveillance Center |
| NACL | Navy-ARPA chemical laser |
| Nanometer | A billionth ($10^{-9}$) of a meter, also written as nm |
| Nanosecond | A billionth ($10^{-9}$) of a second, also written as ns |
| NASP | National aerospace plane |
| NIRF | Neutralizing improvised explosive devices with radio frequency |
| NOP | North Oscura Peak (White Sands Missile Range) |
| NRL | Naval Research Laboratory |
| PA&E | Program analysis and evaluation |
| PB | Particle beam |
| RF | Radio frequency |
| RMA | Revolution in military affairs |
| SAM | Surface-to-air missile |
| SAR | Synthetic aperture radar |
| SBL | Space-based laser |
| SMC | Space and missile command |
| SSHCL | Solid-State Heat Capacity Laser |
| THz | Terahertz ($10^{12}$ hertz, or a thousand billion oscillations per second) |
| TILL | Tracking illuminator laser |
| UNFTP | Unified Navy Field Test Program |
| USA | United States Army |
| USAF | United States Air Force |
| USN | United States Navy |
| WSMR | White Sands Missile Range |

# Bibliography

Abrahamson, James A., and Henry F. Cooper. "What Did We Get for Our $30-Billion Investment in SDI/BMD?" National Institute for Public Policy, 1999.

*Active Denial Technology: Directed Energy Non-Lethal Demonstration.* Air Force Research Laboratory Fact Sheet, Office of Public Affairs, Kirtland AFB, 87117–5776, March 2001.

Adelman, Kenneth L., and Norman R. Augustine. *The Defense Revolution: Intelligent Downsizing of America's Military.* San Francisco: Institute for Contemporary Studies Press, 1990.

"AFRL's Active Denial System to Begin Vehicle-Mounted Testing." *Inside the Air Force,* September 19, 2003.

Ames, Ben. "Non-Lethal Weapons Give Soldiers More Options to Fight Terrorism." *Military and Aerospace Electronics,* August 2003.

Anderson, Jon R. "Nonlethal Weapons Pack Powerful Punch." *European Stars and Stripes,* April 9, 2000.

Arkin, William M. "The Pentagon's Secret Scream." *Los Angeles Times,* March 7, 2004.

———. "Pulling Punches: Big Plans for Futuristic, Nonlethal Weapons Are Afoot, But Their Use Would Raise Troubling Questions." *Los Angeles Times,* January 4, 2004.

———. "'Sci-Fi' Weapons Going to War." *Los Angeles Times,* December 8, 2002.

Babcock, H. W. "The Possibility of Compensating Astronomical Seeing." *Publ. Astron. Soc. Pac.* 65 (1953): 229–236.

Baker, William. Interview by author, April 28, 2004. Baker is chief scientist of the Air Force Research Laboratory's Directed Energy Directorate.

Basov, N. G., and A. N. Oraevskii. "Soviet Physics." *JETP* 17, no. 1171 (1963).

Bates, George. Interview by author, October 18, 2004. Bates is past director of high-power microwave research, PMS–405, NavSea Command, U.S. Navy.

Beason, Charles W. Interview by author, 2001–2005. Beason is past coprogram manager (with Dr. Kirk Hackett) of the Active Denial system and test subject 1 of the first full-body human exposure of Active Denial.

Beason, J. Douglas. *DoD Science and Technology: Strategy for the Post–Cold War Era.* Washington, D.C.: National Defense University Press, 1997.

Bennett, John T. "New Non-Lethal Weapons Being Tested, Might Be Ready Within a Year." *Inside the Pentagon*, July 3, 2003.

Blick, Dennis W., Eleanor R. Adair, William D. Hurt, Clifford J. Sherry, Thomas J. Walters, and James H. Merritt. "Thresholds of Microwave-Evoked Warmth Sensations in Human Skin." *Bioelectromagnetics* 18 (1997): 403–409.

Brinkley, C. Mark. "Senior Officials Have Unique Task in Assessing Weapon's Future." *Marine Corps Times*, July 30, 2001.

_____."Zapper Techniques." *Marine Corps Times*, July 30, 2001.

Broad, William J. "Report Urges U.S. to Increase Its Efforts on Nonlethal Weapons." *New York Times*, November 6, 2002.

Brown, Malina. "DoD's Active Denial Non-Lethal Gun Could Be Fielded As Soon As 2007." Inside Defense.com, March 3, 2003.

Bunker, Robert J., ed. *Nonlethal Weapons: Terms and References.* Institute for National Security Studies Occasional Paper 16. United States Air Force Academy, July 15, 1997.

Bunko, Tracy. "Air Force Approves Initial Production of Laser-Based Jammer Slated for C-17, C-130." Air Force Material Command Public Affairs, AFMC news release 0832, August 23, 2002.

Burger, Kim. "Exotic Non-Lethal Weapons to Quell Mob Rule." *Jane's Defence Weekly*, May 14, 2003.

Canavan, Gregory H. Interview by author, December 21, 2004.

Canavan, Gregory H., Nicholaas Bloembergen, and C. Kumar Patel. "Debate on APS Directed-Energy Weapons Study." *Physics Today*, November 1987, 48–50.

Canavan, Gregory H., and John F. Lilley. "Managing Public Sector Research and Development: Innovation Versus Responsiveness." In *Defense Technology*. Edited by A. Clark IV and J. Lilley. New York: Praeger, 1989.

Castelli, Christopher J. "Navy Looks to Put Invisible Heat Ray Capability on Fleet Vessels." *Inside the Navy*, June 11, 2001.

_____. "Questions Linger About Health Effects Of DOD's 'Non-Lethal Ray.'" *Inside the Navy*, March 26, 2001.

Chalfin, Steven, John A. D'Andrea, Paul D. Comeau, Michael E. Belt, and Donald Hatcher. "Millimeter Wave Absorption in the Nonhuman Primate Eye at 35 GHz and 94 GHz." *Health Physics* 83, no. 1 (2002): 83–90.

Christenson, Sig. "Brooks Dedicates Energy Weapons Lab." *San Antonio Express-News*, June 8, 2001.

Dao, James. "Pentagon Unveils Plans for a New Crowd-Dispersal Weapon." *New York Times*, March 2, 2001.

*Developing Effective Non-Lethal Weapon Options Is Needed to Enhance Naval Force Capabilities.* National Research Council, Committee for an Assessment of Non-Lethal Weapons Science and Technology, November 4, 2002.

*Directed Energy Weapons: Technology, Applications, and Implications.* Lexington Institute, 1600 Wilson Blvd., Arlington, Va., 22209, February 2003.

Duffner, Robert W. *Airborne Laser: Bullets of Light.* New York: Plenum, 1997.

Duffner, Robert W., Robert R. Butts, J. Douglas Beason, and Ronald R. Fogleman. "Directed Energy: The Wave of the Future," in *The Limitless Sky: Air Force Science and Technology Contributions to the Nation,* edited by Alexander H. Lewis. Air Force History and Museums Program, United States Air Force, Washington, D.C., 2004, 213–254.

Fialka, John J. "Group Urges Military to Change Non-Lethal Weapons Research." *Wall Street Journal*, November 5, 2002.

Foster, Kenneth R., John A. D'Andrea, Steven Chalfin, and Donald J. Hatcher. "Thermal Modeling of Millimeter Wave Damage to the Primate Cornea at 35 GHz and 94 GHz." *Health Physics* 84, no. 6 (2003): 764–769.

Friedman, Norman. *Desert Victory: The War for Kuwait.* Annapolis, Md.: Naval Institute Press, 1991.

Gerry, E. T. *IEEE Spectrum* 7, no. 51 (1970).

Gilmore, Gerry J. *DoD Harnesses Technology in Search for Nonlethal Systems.* American Forces Press Service, January 22, 2004.

Glenn, Russell W. "Letting God Rest." *Armed Forces Journal*, May 2003.

Hackett, Kirk. Interview by Bob Duffner, October 26, 2001.

Hearn, Kelly. "New Non-Lethal Energy Weapon Heats Skin." UPI, Washington, February 26, 2001.

Jauchem, James R. "A Literature Review of Medical Side Effects from Radio-Frequency Energy in the Human Environment: Involving Cancer, Tumors, and Problems of the Central Nervous System." *Journal of Microwave Power and Electromagnetic Energy* 38, no. 2 (2003): 103–124.

Jauchem, James R., Kathy L. Ryan, and Melvin R. Frei. "Cardiovascular and Thermal Responses in Rates During 94 GHz Irradiation." *Bioelectromagnetics* 20 (1999): 264–267.

Kaufman, Gail. "Pentagon Program Assures 'Human Effects' of Non-Lethal Weapons Are Reviewed." *Stars and Stripes Omnimedia*, April 2, 2001.

Kehs, Al. Introduction to RF-DEWs. Directed Energy Professional Society short course, Rockville, Md., October 2004.

Knickerbocker, Brad. "The Fuzzy Ethics of Nonlethal Weapons." *Christian Science Monitor*, February 14, 2003.

Kreisher, Otto. "Military Continues Research Into Use of Nonlethal Weaponry." *San Diego Union-Tribune*, December 9, 2002.

Kucera, Joshua. "US Forces Augment Non-Lethal Ability." *Jane's Defence Weekly*, February 4, 2004.

Lamberson, Donald. Interview by author, November 6, 2004.

Lapin, Greg. *N9GL's RF Safety Column: The Military's New RF Weapon*. ARL RF Safety Committee, March 28, 2001.

Loree, Diana. Interview by Barron K. Oder, March 3, 2004. Loree is Active Denial program manager, Directed Energy High-Power Microwave Applications.

Lowe, Christian. "Nonlethal Weapons Are Important, Too." *Air Force Times*, January 26, 2004.

Lumpkin, John J. "Kirtland Weapon Just Hurts." *Albuquerque Journal*, March 1, 2001.

Madey, J. M. J. *Journal of Applied Physics* 42 (1979): 1906.

Mason, Patrick A., Thomas J. Walters, John DiGiovanni, Charles W. Beason, James R. Jauchem, Edward J. Dick Jr., Kavita Mahajan, Steven J. Dusch, Beath A. Shields, James H. Merritt, Michael R. Murphy, and Kathy L. Ryan. "Lack of Effect of 94 GHz Radio Frequency Radiation Exposure in an Animal Model of Skin Carcinogenesis." *Carcinogenesis* 22, no. 10 (2001): 1701–1708.

"Marine Corps Times Reports on Secret Weapon Program." *Army Times*, February 23, 2001.

Mihm, Stephen. "Stench Warfare." *New York Times Magazine*, December 15, 2002.

"Military Research Explores Various Nonlethal Weapons." *Baltimore Sun*, November 23, 2002.

Miller, J. "Advances in Chemical Lasers." In *Proceedings of the International Laser Science Conference*, October 1986.

_____. "Advances in Chemical Lasers." In *Proceedings of the International Conference on LASERS '86*, 1986.

_____. "Chemical Lasers." *TRW QUEST Magazine*, Spring 1980, 2–17.

_____. "Chemical Lasers." In *Proceedings of the Symposium on Lasers and Particle Beams for Fusion and Strategic Defense*, University of Rochester, April 1985, 145–157. Also published in *Special Issue of the Journal of Fusion Energy* 5, no. 1 (1985).

_____. "High Power Hydrogen Fluoride Chemical Lasers: Power Scaling and Beam Quality." In *Proceedings of the International Conference on LASERS '87*, 1987.

_____. "Status of High Energy Chemical Lasers." In *Proceedings of the International Conference on LASERS '85*, 1985.

_____. "Status of High Energy Chemical Lasers." In *High Power and Solid State Lasers*, edited by William W. Simmons, 93–100. Bellingham, Wash.: SPIE, 1986.

Mittleman, D. M., M. Gupta, R. Neelamani, R. G. Baraniuk, J. V. Rudd, and M. Koch. "Recent Advances in Terahertz Imaging." *Applied Physics B Lasers and Optics*, 1999.

Morris, Jefferson. "DOD Planning FY '05 S&T Program for Nonlethal Weapons." *Aerospace Daily*, November 5, 2003.

Nakajima, Shigen. "Japanese Radar Development Prior to 1945." *IEEE Antennas and Propagation*, December 1992.

Nalty, Bernard C., ed. *Winged Shield, Winged Sword: A History of the United States Air Force*. Vol. 1, *1907–1950*. Washington, D.C.: Government Printing Office, 1997.

Nelson, D. A., M. T. Nelson, T. J. Walters, and P. A. Mason. "Skin Heating Effects of Millimeter-Wave Irradiation: Thermal Modeling Results." *IEEE Transactions on Microwave Theory and Techniques*," November 2000, 2111–2120.

Nelson, D. A., T. J. Walters, K. L. Ryan, K. B. Emerton, W. D. Hurt, J. M. Ziriax, L. R. Johnson, and P. A. Mason. "Inter-Species Extrapolation of Skin Heating Resulting from Millimeter Wave Irradiation: Modeling and Experimental Results." *Health Physics*, May 2003, 608–615.

"New Technology Drives Away Adversaries." Joint USMC-USAF press release, USMC no. 2001–09, DE release no. 2001–09, February 22, 2001.

Office of Technology Policy, Bureau of Economic Analysis, U.S. Department of Commerce. *US Corporate R&D Investment, 1994–2000: Final Estimates*. Washington, D.C.: Government Printing Office, 2002.

O'Shea, Patrick G., and Henry P. Freund. "Free-Electron Lasers: Status and Applications." *Science*, June 8, 2001, 1853–1858.

Pakhomov, Andrei G., Yahya Akyel, Olga N. Pakhomova, Bruce E. Stuck, and Michael R. Murphy. "Review Article: Current State and Implications of Research on Biological Effects of Millimeter Waves: A Review of the Literature." *Bioelectromagnetics* 19 (1998): 393–413.

Patel, C., and N. Kumar. "Selective Excitation Through Vibrational Energy Transfer and Optical Laser Action in $N_2$-$CO_2$." *Physical Review Letters* 13 (1964): 617–619.

Pogue, Ed. Interview by author, October 18, 2004.

*Policy for Non-Lethal Weapons*. Department of Defense Directive 3000.3, July 9, 1996.

Pringle, Rodney L. "Directed Energy Weapons Roadmap Could Lead to New Applications." *Aviation Week's NetDefense*, March 11, 2004.

Rees, Elizabeth. "Combatant Commands to Be Tasked with Identifying Non-Lethal Needs." *Inside the Air Force*, November 7, 2003.

Reppert, Barton. "Force Without Fatalities." *Government Executive Magazine*, May 1, 2001.

Rogers, Mark E. *Lasers in Space: Technological Options for Enhancing U.S. Military Capabilities*. Occasional Paper no. 2. Center for Strategy and Technology, Air War College, Maxwell Air Force Base, Ala., November 1997.

Rosenberg, Barbara Hutch, and Mark L. Wheelis. "'Nonlethal' Weapons Put Humanity At Risk." *Los Angeles Times*, December 1, 2002.

Rumsfeld, Donald, and Richard Myers. DoD news briefing transcript, August 9, 2002.

Ryan, Kathy L., John A. D'Andrea, James R. Jauchem, and Patrick A. Mason. "Radio Frequency Radiation of Millimeter Wave Length: Potential Occupational Safety Issues Relating to Surface Heating." *Health Physics*, February 2000, 170–181.

Scott, William B. "Navy Accelerates Transition of Technology to Weapons Systems." *Aviation Week and Space Technology*, May 7, 2001.

Sheehan, John J. "Non-Lethal Weapons: Let's Make It Happen." Speech before the Non-Lethal Weapons Conference II, Washington D.C., March 7, 1996.

Sirak, Michael. "US Looks to Build on Non-Lethal Weapon Support." *Jane's Defence Weekly*, November 12, 2003.

Sirak, Michael, and Kim Burger. "The Promise and the Peril of Non-Lethal Weapons." *Jane's Defence Weekly*, August 6, 2003.

Tuttle, Rich. "Large Aircraft Infrared Countermeasures System." *Aerospace Daily and Defense Report,* 2004.

*United States Strategic Bombing Survey: Summary Report (Pacific War) 1 July 1946.* Washington, D.C.: Government Printing Office, 1946.

Vorenberg, Sue. "N.M. Labs, Companies into Scare Tactics: Fright Is Might Using Nonlethal Arms." *Albuquerque Tribune*, January 30, 2003.

Walters, Thomas J., Kathy L. Ryan, David A. Nelson, Dennis W. Blick, and Patrick A. Mason. "Effects of Blood Flow on Skin Heating Induced by Millimeter Wave Irradiation in Humans." *Health Physics*, February 2004, 115–120.

Wissler, John B. "Organization of the Joint Technology Office." *Program Manager*, November–December 2002.

Wodele, Greta. "Defense Officials to Boost Budget for Nonlethal Weapons." *National Journal's Technology Daily*, January 30, 2004.

*ZEUS-HLONS, HMMWV Laser Ordinance Neutralization System.* United States Army Space and Missile Defense Command, Space and Missile Technical Center, Public Affairs Office no. 0183, February 2004.

Zitner, Aaron. "'Nonlethal' Weapons Vital, Panel Says." *Los Angeles Times*, November 5, 2002.

# Index